Before Sudoku

Image on the Title Page

A magic square design over the background of Benjamin Franklin's perfect 16×16 magic square. This design, which was originated by the authors, gives the magic constant of 2,056 when the numbers along the vertical and horizontal wavy white lines are added. In the Franklin magic square background, the numbers in the 16 horizontal rows, the 16 vertical columns, as well as the two corner-to-corner diagonals each add to give the magic constant of 2,056. This and the square's many other unusual properties, to be presented in the book, represent the magic of magic squares.

Before Sudoku

The World of Magic Squares

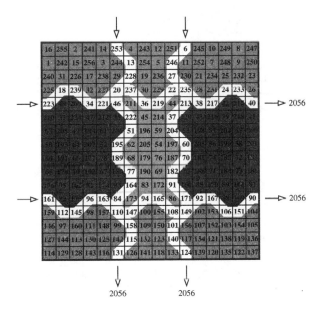

2056

2056

2056   2056

Seymour S. Block
Santiago A. Tavares

OXFORD
UNIVERSITY PRESS
2009

OXFORD
UNIVERSITY PRESS

Oxford University Press, Inc., publishes works that further
Oxford University's objective of excellence
in research, scholarship, and education.

Oxford   New York
Auckland   Cape Town   Dar es Salaam   Hong Kong   Karachi
Kuala Lumpur   Madrid   Melbourne   Mexico City   Nairobi
New Delhi   Shanghai   Taipei   Toronto
With offices in
Argentina   Austria   Brazil   Chile   Czech Republic   France   Greece
Guatemala   Hungary   Italy   Japan   Poland   Portugal   Singapore
South Korea   Switzerland   Thailand   Turkey   Ukraine   Vietnam

Published by Oxford University Press, Inc.
198 Madison Avenue, New York, New York 10016

www.oup.com

Oxford is a registered trademark of Oxford University Press

Library of Congress Cataloging-in-Publication Data
Block, Seymour S.
Before Sudoku: the world of magic
squares / Seymour S. Block and Santiago A. Tavares.
p. cm.
Includes bibliographical references and index.
ISBN 978-0-19-536790-4
1. Magic squares. 2. Sudoku. 3. Mathematical recreations.
I. Tavares, Santiago Alves.   II. Title.
QA165.B56 2009
793.74—dc22   2008018537

1 3 5 7 9 8 6 4 2

Printed in the United States of America
on acid-free paper

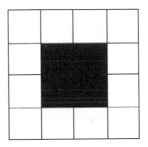

Acknowledgments

This book touches on many unusual subjects and the authors were fortunate to have had the assistance of very knowledgeable people who kindly provided us with valuable information in their various areas of expertise.

We express our appreciation to Roy Goodman and Valerie Ann Lutz of the American Philosophical Society Library for their help in opening up sources of early American literature pertaining to Benjamin Franklin and magic squares. We want to thank Dr. Mark Yang of the Statistics Department of the University of Florida for his help in making important calculations for us. In the area of music, we are indebted to Jim Hale, Paul Richards, and Jim Sain of the School of Music at this university, and not to forget the good work of Robina Cornwell of the Music Library in finding composers whose work related to magic squares. We are grateful to Ellen Cohn and Kate Ohno for pertinent information in the "Papers of Benjamin Franklin" at Yale. Eduard Sole gave us permission to use pictures of the Sagrada Familia Cathedral in Barcelona. Jim Bosworth helped us in locating and presenting photographs used in the book. The people at the Interlibrary Loan Department should be mentioned for they were most helpful in enabling us to obtain difficult-to-find publications that were invaluable in preparing the manuscript.

Harvey Heinz made valuable contributions by granting us permission to use material from his and John Hendricks's book, *Magic Square Lexicon: Illustrated*. Heinz also provided us, on the Internet, additional valuable information on magic squares and their literature. A captivating book on the subject, namely, Clifford Pickover's *The Zen of Magic Squares, Circles, and Stars*, presents a wide range of interesting and artistic material. The author generously permitted us to use some of the material from his book.

We must acknowledge our debt to Benjamin Franklin for having introduced us to the fascinating subject of magic squares. One of us, having written two books about Franklin, became intrigued with his work on that puzzle. But as we delved further into the fascinating topic of magic squares, its long history back to ancient times, its relationship to Sudoku, its use in statistics, its application in art, and music, etc., there arose the inspiration that eventually led to this book.

The authors appreciate the kind cooperation they have received in the preparation of this book by the people at Oxford University Press: Michael Penn, Executive Editor; Edward Sears, Editorial Assistant; Christine Dahlin, Production Editor; and Justin Hargett, Publicity Associate.

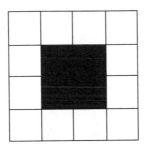

Contents

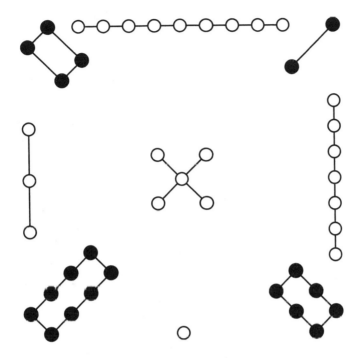

The First Magic Square

The Lo Shu Chinese magic square, estimated 2250 B.C.

Before Sudoku

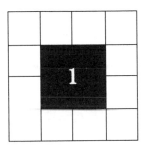

Introduction

Magic squares first appeared in ancient Chinese literature, more than 4,000 years ago. They traveled slowly westward through India, Arabia, and Europe, to America. They captured the attention of mystics, astrologers, clergymen, and some of the world's most brilliant thinkers, including great mathematicians, as well as Benjamin Franklin and Charles Babbage, the inventor of the early computer.

Today we have a very popular offshoot of the magic square, the Sudoku puzzle. This book is written for people like those who do the Sudoku puzzle, average readers, for it contains only simple math. It explores magic squares in their many realms: in ancient and medieval history, in art and design, in music, and in practical applications in statistics and electronics. It also takes us on a space probe of Mars and, even further, on a trip to the fourth dimension of space.

This is a puzzling book; it deals in puzzles, number puzzles. From earliest times, people have been fascinated by numbers, and this book takes us everywhere numbers have taken humankind. This is not a book for mathematicians; it is a book for everybody who has an ever-itching curiosity along with an insatiable imagination.

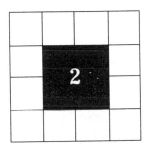

Magic Squares and Sudoku

Figure 2.1 is a magic square. It is like a crossword puzzle filled with numbers instead of letters. This is a 4 × 4 square, but like a crossword puzzle it can be any size from 3 × 3 and up. What makes it magic is that when you add the numbers in any line across or any column down, you get the same sum, namely, 34. In this case, the sum of each of the diagonals (16, 10, 7, 1) and (4, 6, 11, 13) is also 34, which makes it a special or perfect magic square. According to tradition, the numbers begin with 1 and fill the little squares, or cells, so a 4 × 4 square contains all the numbers, without repetition, from 1 to 16. It may be surprising, but there are 880 different arrangements

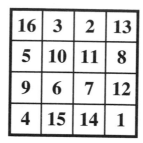

Figure 2.1   Benjamin Franklin 4 × 4 magic square.

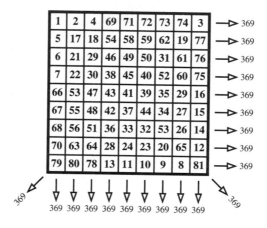

Figure 2.2   Magic square 9 × 9.

of the numbers from 1 to 16 in a 4 × 4 square that yield the magic sum of 34. Try composing ten, or perhaps just one. It isn't easy. Consider that there are over 2.6 trillion different possible arrangements of 16 numbers to fill the 4 × 4 square and 10^{88} (which in miles represents the distance to the edge of the universe) possible arrangements to fill an 8 × 8 square [16].

Figure 2.2 shows a 9 × 9 magic square (Wilke [116]), and Figure 2.3 shows a 9 × 9 completed square of the popular, well-known Sudoku puzzle as it appeared in a newspaper. In Figure 2.2,

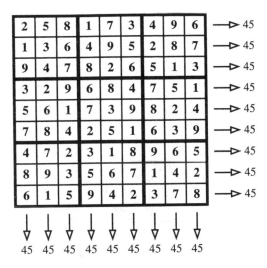

Figure 2.3   Completed 9 × 9 Sudoku

each row adds up to give the same number for that square, so both squares are magic squares. The square in Figure 2.2 differs from the Sudoku (Figure 2.3), which does not complete the number sequence of 1 to 81 ($9^2 = 81$) as in Figure 2.2. The Sudoku is divided into three parts and the numbers in each row and column include 1 to 9, also in the nine 3 × 3 subsquares. Whereas in a magic square (Figure 2.2) no number may be repeated, in the Sudoku (Figure 2.3) numbers are repeated in each row and column but may not be repeated within individual rows or columns.

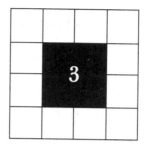

History of Sudoku

The Sudoku puzzle is derived from an offshoot of the magic square, called the Latin square, that was created by the great mathematician Leonhard Euler in 1783. The Sudoku was the invention of an American architect from Indianapolis, Howard Garns, in 1979. It appeared in the Dell puzzle magazine where it was called Number Place, and it is still called that today in that magazine. The originator was determined by some detective work by Will Shortz, the puzzle editor of the *New York Times*. Nobody at the magazine now was there in 1979 or knew what had occurred then, and the magazine had not attached Garns's name to the puzzles. Shortz examined the list of contributors in issues when the Number Place puzzle appeared and found that the name Howard Garns was listed with the contributors but was absent whenever that puzzle did not appear. He was able to confirm his evidence with Garns's associates at his architectural firm who witnessed him creating the puzzles. Unfortunately, Garns died in 1989 before his puzzle had achieved its worldwide popularity (Hayes [18]).

Garns had taken Euler's Latin square and applied it to a 9 × 9 grid and added nine 3 × 3 subgrids, each with the numbers 1 to 9. The scene now shifts to Japan when in 1984 the Japanese puzzle publishing firm Nikoli discovered this puzzle and included a similar one in one of their magazines. It quickly became popular in Japan under the name *Suji wa dokushin ni kagiru*, meaning "the numbers must be single." That name was shortened to Su-doku or "single numbers" and, following some improvements, Sudoku became the best-selling puzzle in Japan. It is reported that now there are more than 600,000 copies of different magazines containing a Sudoku puzzle published in Japan every month. The name Sudoku is trademarked in Japan but not in other countries, yet ironically, in Japan many know it by its English name, Number Place, and in English-speaking countries the name Sudoku is preferred.

The president and founder of Nikoli, Maki Kaji, who launched Sudoku and gave it its name, is an interesting fellow. He is a college dropout who worked as a waiter and a construction worker, and then started a small publishing business. He was not a puzzle fancier but started a puzzle magazine because there weren't any in Japan at that time. Looking for new puzzles for his magazine, he saw Number Place in the Dell magazine in 1984, and after rearranging the numbers and grids he published it in Japan, where it remained for the next ten years until it was discovered by a foreigner and became known to the rest of the world. Number puzzles have a long history in Japan because the Japanese love to challenge themselves in their spare time. Kaji thinks Japan is therefore a good incubator of puzzles. He has a small crew who edit and refine the 1,000 puzzles that are sent in by readers every month, just as was done with

Number Place in its conversion to Sudoku. Kanji says he had no idea Sudoku would be such a big success. He just keeps trying with new puzzles and has had two recent successes in Japan, Kakuro and Slitherlink, that he has tried to launch abroad.

Despite Sudoku's worldwide fame, Kanji has not made a financial killing because his trademark is only registered in Japan, but that doesn't bother him. He prefers to remain small. Big companies have tried to buy out Nikoli, but Kanji refused all offers, saying, "I like the idea of a home cake shop that runs out at 3 PM and then we all just go home." He has never advertised and says the absence of advertising and marketing means freedom. "I am the rulebook, what I say today may change tomorrow. I don't go with the flow—I flow."

Kanji abhors computer-produced puzzles. He says, "Our puzzles are hand made, not computer-generated like most of the Sudoku found overseas. That kind lacks a vital ingredient that makes puzzles enjoyable—the sense of communication between solver and author." He says his puzzles aren't stressful; they are absorbing but never boring.

One reason Kanji cherishes his freedom is his love for the race track, where he goes to get excitement and forget his troubles. In fact, that is where he got the name of his company, Nikoli. It was the name of a race horse that he liked, the Epsom Derby favorite. "Being at the racetrack," he says, "is the same as working puzzles. At the track you see old guys screaming like children, they're really into it. With puzzles too, it's the excitement not knowing how it's going to end" [101, 102].

The scene shifts again, this time to Hong Kong where Wayne Gould, a New Zealander, was a judge in British-run Hong Kong prior to 1977. He discovered Sudoku on a visit to Japan and persuaded the *Times* of London in 2004 to print some of the Sudoku puzzles he had generated. The fad hit like a bomb burst, affecting Australia and New Zealand as well. Some 260 newspapers rushed to follow the *Times*, with radio and TV joining the crowd. Sky One

Figure 3.1   Aerial photo of world's largest Sudoku—a 275-foot-square puzzle. Note: There is a truck at the top left of the puzzle.

outdid them all with the world's largest Sudoku, a 275-foot-square puzzle carved on the side of a hill near Bristol, England (Figure 3.1).

This was considered a significant piece of public art. Up to a million cars were said to have passed the site in seven days, and it was recognized in the *Guinness Book of World Records*. A prize of £5,000 ($10,000) was offered to the winner who first solved the puzzle, but it was discovered that this puzzle had more than one solution, so all who had correct solutions were promised to receive the same prize [111]. A book on how to do Sudoku became a national best-seller, and the government-sponsored teachers' magazine recommended Sudoku to be used as an exercise for the brain in classrooms and also to treat people with disorders such as Alzheimer's.

Finally, in 2005 Sudoku completed its worldwide circuit and returned to its place of origin, America, where it gained instant popularity, appearing in newspapers and magazines. Today the puzzle has spread to at least 70 countries, over 600 newspapers, with clubs, online chat rooms, videos, card games, competitions, and many books. Sudoku puzzles vary in difficulty from easy to difficult to fiendish, diabolic, or evil. There are many varieties of puzzles, with the 9 × 9 grid with 3 × 3 boxes being the most common. But there are 4 × 4 grids with 2 × 2 boxes, 6 × 6 grids with 2 × 3 boxes, and so on. The U.S. championship in 2005 had a Sudoku with parallelogram regions that wrapped around the outer border of the puzzle as if the grid were toroidal. Dell publishes Number Place puzzles with 16 × 16 grids, and Nikoli produces 25 × 25 Sudoku giants. There are also alphabetical variants that use letters that spell something instead of using numbers, and many others that are quite bizarre [77, 113].

The first world Sudoku championship was held in Lucca, Italy, in 2006 with participants from 22 countries represented. Jana Tylova, an economist from the Czech Republic, was the winner. A runner-up was a Harvard chemistry graduate student, Thomas Snyder. There were two world championships in 2007: one in Prague on March 31 and another in Rio de Janeiro on October 10. Thomas Snyder won first place in both of these. On October 20, 2007, a national championship was held in Philadelphia with 900 contestants and 150 observers. Thomas Snyder was there as well and took first place, winning $10,000 and a trip to India for the third world championship.

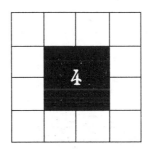

Some Techniques for Solving Sudoku Puzzles

4.1 What Is Sudoku?

The Sudoku or Number Place, is a Latin square divided into subsquares (boxes) as shown in Figure 4.1.

4.2 Solving the Sudoku

The Playing Field

Figure 4.2 shows the playing field for the Sudoku. It is a 9×9 square with nine rows marked $A, B, C, \ldots, I$; nine columns marked $a, b, c, \ldots, i$; and nine boxes marked 1 to 9 (Figure 4.3).

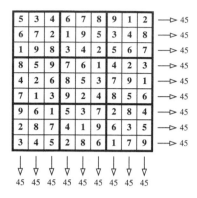

Figure 4.1 A 9 × 9 Sudoku square (solved Sudoku puzzle).

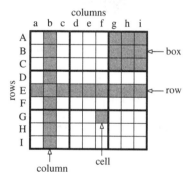

Figure 4.2 A grid with 81 cells, nine rows, nine columns, and nine boxes.

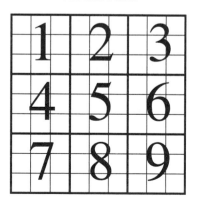

Figure 4.3 Enumeration of the boxes.

	a	b	c	d	e	f	g	h	i
A	8	9			4	6	3		1
B	4	6				5		8	9
C			5				4		
D	2	3	9		5			6	4
E			4		3	9		1	
F				6	2		5		
G	5		1	9	7	2			8
H		8	3			1	9	4	2
I				4	8			7	

Figure 4.4   A Sudoku sample puzzle.

Figure 4.4 is a sample puzzle. Observe that some of the cells have numbers and some do not. The numbers that are given will be different in every puzzle. Like in a crossword puzzle, you must fill in the empty cells with the correct numbers. How do you find the correct numbers? You will know you have the correct numbers when each box, each row, and each column is filled with numbers 1 to 9 (Figure 2.3 or Figure 4.1).

How to Proceed

1. Starting with box 1, note the missing numbers are 1, 2, 3, 7.

2. Go to the empty cell in row A column c. We know that the rows, columns, and boxes must have numbers 1 to 9, which means no repetitions are possible. Row A and column c both contain 1 and 3 but no 2 or 7, so we put 2 and 7 in cell Ac, as in Figure 4.5.

3. Proceeding to cell Bc, we observe that 1 and 3 are in column c and cannot be repeated in that column, but there is no 2 or 7 in column c or row B, therefore those numbers are placed in Bc.

	a	b	c	d	e	f	g	h	i
A	8	9	2, 7	2, 7	4	6	3	2, 5	1
B	4	6	2, 7	1, 2 3, 7	1	5	2, 7	8	9
C	1, 3 7	1, 2 7	5	1, 2 3,7,8	9	3, 7 8	4	2	6, 7
D	2	3	9	1, 7 8	5	7, 8	7, 8	6	4
E	6, 7	5, 7	4	7, 8	3	9	2, 7 8	1	7
F	1, 7	1, 7	7, 8	6	2	4, 7 8	5	3, 9	3, 7
G	5	4	1	9	7	2	6	3	8
H	6, 7	8	3	5	6	1	9	4	2
I	6, 9	2	6	4	8	3	1	7	5

Figure 4.5   A puzzle, step 1.

4. Now to cell *Ca*. Row C has no 1, 2, 3, 7, but column *a* has a 2, so we put 1, 3, 7 in *Ca*.

5. For cell *Cb*, row C has no 1, 2, 3, 7, but column *b* has a 3, thus we put 1, 2, 7 in *Cb*.

6. Move forward to box 2, which lacks 1, 2, 3, 7, 8, 9. Going to *Ad*, row A contains 1, 3, 8, 9 and column *d* contains no help, so 2 and 7 are left to go in *Ad*.

7. The next empty cell in box 2 is *Bd*. Here row B can supply 8, 9, but column *d* can offer no other needed numbers, so we must write 1, 2, 3, 7 in *Bd*.

8. For *Be*, row B can furnish 8, 9, and column *e* provides 2, 3, 7, 8 of the missing numbers, leaving only 1 for the empty cell. This provides us with the first number we are seeking that is equivalent to the given numbers.

9. Next to cell *Cd*: Row C gives us no help, and column *d* supplies only 9. Therefore *Cd* is filled with 1, 2, 3, 7, 8.

10. On to *Ce*: Row C is no help, but column *e* provides 2, 3, 7, 8, and because we have 1 in *Be*, this leaves only 9 to inhabit cell *Ce* as our second sole occupant.

11. The last empty cell in box 2 is *Cf*. Row C can furnish no missing numbers, but column *F* gives us 1, 2, and 9, leaving 3, 7, 8 to put into *Cf*.

We proceed in the same way in the remaining boxes to give us the results shown in Figure 4.5.

Basic Elimination Procedure

To solve this puzzle we need just one number in each cell, so our next job is to eliminate the excess numbers in the cells in Figure 4.5.

Because there can be only numbers 1 to 9 in each box, each row, and each column, we can start with the newly found sole occupants, like 1 in *Be* and 9 in *Ce* and eliminate other 1s and 9s in their boxes, rows, and columns. When this process has proceeded for all the other sole occupants, new sole occupants will be uncovered, and the process can be repeated again, until all the multiple cell occupants have been eliminated. If there have been no mistakes, a check of the boxes, rows, and columns will show that each contains the required sequence of 1 to 9. In Figure 4.6, the numbers not eliminated are indicated inside a circle. It is a

	a	b	c	d	e	f	g	h	i
A	8	9	2⑦	②7	4	6	3	2⑤	1
B	4	6	②7	1 2 ③7	①	5	2⑦	8	9
C	1③ 7	①2 7	5	1 2 3⑦8	⑨	3 7 ⑧	4	②	⑥7
D	2	3	9	①7 8	5	1⑦ 8	7⑧	6	4
E	⑥7 8	⑤7	4	7⑧	3	9	②7 8	1	⑦
F	①7 8	1⑦	7⑧	6	2	④7 8	5	3⑨	③7
G	5	④	1	9	7	2	⑥	③	8
H	6⑦	8	3	⑤	⑥	1	9	4	2
I	6⑨	②	⑥	4	8	③	①	7	⑤

Figure 4.6   A puzzle, step 2.

	a	b	c	d	e	f	g	h	i
A	**8**	**9**	7	2	**4**	**6**	**3**	5	**1**
B	**4**	**6**	2	3	1	**5**	7	**8**	**9**
C	3	1	**5**	7	9	8	**4**	2	6
D	**2**	**3**	**9**	1	**5**	7	8	**6**	**4**
E	6	5	**4**	8	**3**	**9**	2	**1**	7
F	1	7	8	**6**	**2**	4	**5**	9	3
G	**5**	4	**1**	**9**	**7**	**2**	6	3	**8**
H	7	**8**	**3**	5	6	**1**	**9**	**4**	**2**
I	9	2	6	**4**	**8**	3	1	**7**	5

Figure 4.7   Solution to Puzzle 4.4.

good idea to check your work as you go along because it is easy to overlook a number here and there, and it will save you time and frustration if you are careful.

After the numbers outside the circles have been eliminated, the completed puzzle looks like Figure 4.7.

Examine to see if all rows, columns, and boxes contain numbers 1 to 9.

4.3   The Naked Double

The naked double is a useful tool for eliminating excess numbers in the cells. After penciling in all the possible numbers in the cells in any puzzle, you will have a grid that will be similar to Figure 4.5. If we consider Figure 4.5, we notice that row A has two cells, Ac and Ad, that have numbers 2 and 7. These are called a naked double because they have just two numbers and no more, and because they appear together at least twice in a row, column, or box. Row B has 2 and 7 in Bc and Bg. Column a has a naked double of 6, 7 in Ea and Ha. Column c has a naked double of 2, 7 in Ac and Bc. This

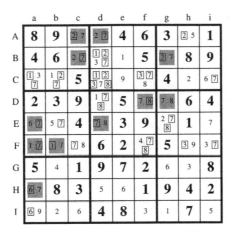

Figure 4.8 Solution to Puzzle 4.4.

pair is also in box 1, so it is useful in eliminating numbers in both the column and the box. In Figure 4.8, all the naked pairs are in gray cells.

Now for the elimination process. In row A we know that the sole occupant of Ac or Ad must be a 2 or a 7, so we may eliminate all other 2s or 7s in that row. That happens to be the 2 in Ah. The elimination is indicated by a rectangle around the number. In column a the pair of 6, 7 allows us to strike out the 7 in Ca, the 7 in Fa, and the 6 in Ia. In column c the 2, 7 pair permits us to erase the 7 in Fc, and because this pair is in box 1, we can delete the 7s in Ca and Cb in addition to the 2 in Cb. So you see, the naked double is a helpful tool that will allow us to save time in solving the puzzle. Figure 4.8 shows the cancellation of the excess numbers by this procedure.

4.4   Another Option

It may be more efficient to proceed by filling boxes that contain the larger number of given numbers and are adjacent to boxes that are also well filled. For example, in Figure 4.4, on page 13, box 8 has only three empty cells and is adjacent to box 9 with four empty cells, box 5 with four empty cells, and box 7 with five. Start with box 8, which is missing 3, 5, 6. Number 3 can't go in row *H* so it must go in *If*. Number 5 can't go in column *e* so it finds itself in *Hd*. That leaves *He* for 6. Noting that row *H* has only one empty cell and lacks a 7, place it in *Ha*.

Proceed to box 9, which lacks 1, 3, 5, 6. Number 1 cannot go in row *G* or column *i*, leaving only the cell *Ig* to place 1. Three cannot go in row *H* or column *g* so it must go in cell *Gh*. Five cannot go in row *G* or column *g* so it must fall into cell *Ii*, leaving cell *Gg* for 6. In box 7, row *G* has only one cell vacant, missing a 4. A 4 is forbidden in row *H* or *I* so it must be put in cell *Gb*. Staying with box 7, it requires 2, 6, 9 to fill the voids in row *I*. Trying 2, it can't go in column *a* so we must chalk it in cells *Ib* and *Ic*. Six cannot go in *Ib* so we have to try *Ia* and *Ic*. Nine cannot fill *Ib* or *Ic* so slip it into *Ia*. Circle 9 in *Ia* and circle 6 in *Ic*. Now we have single occupants in the cells of boxes 7, 8, 9. Column *e* needs only 9 and 1 to fill its two remaining empty cells. The 1 can fit in cells

	a	b	c	d	e	f	g	h	i
A	8	9	⑦	②	4	6	3	2⑤	1
B	4	6	②	③	①	5	⑦	8	9
C	1③	①	5	3⑦	⑨	⑧	4	②	⑥
D	2	3	9	①	5	⑦8	7⑧	6	4
E	⑥	⑤	4	7⑧	3	9	②	1	⑦
F	①	1⑦	⑧	6	2	④	5	⑨	③
G	5	④	1	9	7	2	⑥	③	8
H	⑦	8	3	⑤	⑥	1	9	4	2
I	6⑨	②	2⑥	4	8	③	①	7	⑤

Figure 4.9 Option 2 for solving Puzzle 4.4.

Be or *Ce* but 9 cannot go into *Be* so it goes into *Ce* and 1 goes into *Be*, completing that column.

The rest of the puzzle is routine and the results are shown in Figure 4.9.

Next Step

The process here will provide you with the skills necessary to solve all the easy Sudoku puzzles that appear in the newspapers and magazines as well as those labeled moderate. Books of Sudoku puzzles may distinguish between easy, moderate, and difficult puzzles. When you feel you are ready to attack the difficult ones, you will find hints in those books on how to go about it. In the meantime, try your hand on the puzzles in Figure 13.6 and Figure 13.7. You will find the solution to them in Figure 16.18 and Figure 16.19.

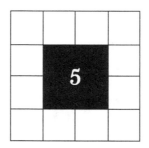

History of Magic Squares

5.1   Magic Squares 4,000 Years Ago: China

Magic squares go back in ancient history, before the Romans, before the Greeks, before the Hebrews, to the ancient Chinese. Over 4,000 years ago, 2200–3000 B.C., the mythical Emperor Yu is said to have observed some curious markings on the undershell of a large tortoise that had emerged from the Lo River. The marks had the shape of a square, and when they were counted either horizontally, vertically, or diagonally they produced the same number. This was interpreted as a mystical or magical phenomenon and thus these mathematical structures have become known as magic squares. The first written reference to the magic square appeared in 300 B.C. in the writings of Zhuang Zi [69].

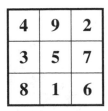

Figure 5.1   The Lo Shu magic square shown with conventional numbers.

Throughout history up until modern times, magic squares were considered to possess magical or supernatural properties and were used as charms to dispel evil and bring good fortune. Today, they are regarded with fascination but without the mysterious, superstitious properties.

A depiction of the magic square supposedly observed by Emperor Yu is shown as the figure following the Contents and with conventional numbers in Figure 5.1. In this figure, the magic number is 15. That is, the sum of the numbers in each of the three horizontal rows, each of the three vertical columns, and each of the two diagonals is 15. The magic number is also referred to as the magic sum or magic constant. This square is referred to as the Lo Shu because the sacred tortoise with the magic markings came from the Lo River (a branch of the Yellow River), and Shu refers to the marks or writing on it.

The Chinese regarded the magic square with awe that did not favor the development of the mathematical advancement of the subject. They attributed some of the numbers to male, or active, characteristics (yang) and others to female, or passive, characteristics (yin), as seen in Figure 5.2. In the Lo Shu in the front of the book, the dots in black depict the yang numbers and those in white the yin numbers. The numbers in the square were believed to be related to basic properties of the Earth, and indeed to the whole order of the universe (Figure 5.3). They show the yang and yin properties, as indicated by the shading within the box.

The Lo Shu square is known as a magic square of *order* 3, that is, the square is made up of three numbers in the horizontal rows and three numbers in the vertical columns.

Yang		Yin
Active		Passive
Hot		Cold
Life		Death
Summer		Winter
Male		Female
Day		Night
Odd		Even
Sun		Moon
Fire		Water

Figure 5.2   The yang and yin ancient Chinese symbols.

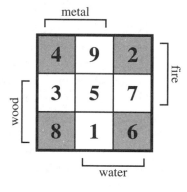

Figure 5.3   The Lo Shu with ancient symbols.

5.2   India and the Middle East

By the first century A.D. a magic square of order 4 was reported in India. During the first century, other order 4 magic squares were reported in India, and the scene for magic square activity had shifted from China to India, and then, following eastward, to Arabia and Asia Minor ([8], [9]). Under Arab influence, we

ד	ט	ב
ג	ה	ז
ח	א	ו

4	9	2
3	5	7
8	1	6

Figure 5.4   Order 3 magic square in Hebrew (letters) numbers.

find the Jewish philosopher-astrologer Abraham ibn Ezra (1090–1167) using the Lo Shu numbers with Hebrew letter-numbers (Figure 5.4 [68]).

Astrology, which associated the motion of the heavenly bodies with human activities, was prevalent in India, Arabia, and Europe during the Middle Ages. It aroused great interest in magic squares. Thus, the scholar Al-Buni, a Muslim in North Africa in the thirteenth century, published a description of amulets worn for protection against evil spirits in which the planet Saturn was represented by a third-order square, Jupiter a fourth-order square, Mars a fifth-order, the Sun a sixth-order, Venus a seventh-order, Mercury an eighth-order, and the Moon a ninth-order square ([8]). In those days the Sun and the Moon were considered to be planets. We learn from the astrologers that the planets each had their own personality and each affected our lives and well-being; thus, in their understanding the relationship of magic squares to the planets was of great importance.

It should be noted that it was not only the Muslims and the Hindus who accepted astrology during the Middle Ages. Christians and Jews in Europe also believed in the interpretation by astrologers through magic squares in the effect of the planets, as shown by the amulets that were produced containing Christian words, Hebrew letters, and magic squares (Figures 5.5 and 5.6 [62]).

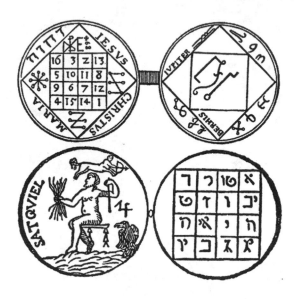

Figure 5.5   Jupiter amulet, magic incantation.

Figure 5.6   Mars amulet.

5.3   Europe

Magic squares made their appearance in Europe through Asia Minor and Constantinople (now Istanbul in modern Turkey). Manual Moschopolous, a Greek-Byzantine scholar, wrote a book on magic squares around 1300 based on the work of Al-Buni. His work was different from preceding writings in that he treated magic squares in a logical mathematical way instead of the mystical manner of earlier writers. This should have changed all further work, but it didn't because it was not widely disseminated and was only discovered 300 years later in Paris, according to Frank Swetz in his book *Legacy of the Luoshu* [72].

In Europe, as in the East, magic squares continued to be associated with the great mysteries of the universe interpreted through astrology. German scholar Henry Cornelius Agrippa (1488–1535) brought us the greatest collection of beliefs and practices during the medieval period through his books on astrology, divination, and magic, the most famous being *The Philosophy of the Occult*. It included magic squares of orders 3 to 9 and was of great influence in Europe. In 1514, the famous Renaissance sculptor-artist Albrecht Dürer produced a wood engraving titled *Melencolia I* that includes a magic square of order 4. It features a brooding alchemist in his shop sitting with his head on his hand (Figure 5.7 [83]). He is sullen, all his instruments are in disarray, nothing is being weighed on the balance, nobody is on the ladder, the bell is not ringing, and time is running out in the hourglass. All of these symbols are demonstrating the effect of the alchemist's melancholy [84].

On the wall above the alchemist's head is a magic square (Figure 5.8). What makes this square unusual is that Dürer has been able to work in the year 1514 in the bottom row of the square, which is the year of the completion of the work. The Babylonians, who in their cosmology assigned different heavenly bodies to magic squares, also assigned special human properties to each of

Figure 5.7   Dürer's medieval engraving *Melencolia I.*

Figure 5.8   Melencolia I magic square, including the year 1514.

them [85]. Thus, Saturn, the square for order 4, was given sanguine and salutary properties, and following astrology, the artist was using this magic square to bring the alchemist out of his melancholy.

Despite the persistence of age-old beliefs, the Renaissance was beginning to take hold in Europe. In Germany we see its effect on the work in the magic squares of Michael Stifel (1489–1559). Stifel was a graduate of the University of Wittenberg who entered the monastery and became a Catholic priest. As a mathematician, we know Stifel best for giving us the plus and minus signs, + and −. After meeting Martin Luther, Stifel became a Lutheran pastor, and Luther found him a parish in the town of Lochau. Regarding magic squares, he produced squares of order 5, 9, and 16 and bordered (nested, concentric) squares. But using his math in religious matters got him into trouble. First, he calculated that the world was going to end exactly on October 19, 1533, at 8 A.M., and he announced this to all the folks in Lochau. They immediately stopped farming and working, divesting themselves of material possessions so as to be ready to go to heaven on that day. When the fatal day came and the end hadn't arrived, the poor pastor had to flee for his life. Fortunately, Luther found him another parish, but he still wasn't free of trouble. Using new calculations based on the name of the pope, he determined that Leo X, the pope, was really the devil, the Antichrist in disguise. Again Luther was able to save him [67].

In nearby France in the next century, Bernard Frénicle de Bessey (1605–1675) was an amateur mathematician who corresponded with famous men like Descartes, Fermat, and Huygens and is best known today for his contributions to number theory. We know him for his work on the magic square of order 4. Frénicle produced 880 different magic squares of order 4, all that have been shown possible. The one Dürer put in his *Melencolia I* was one of them, which received the number 175 in Frénicle's list of the 880 [12].

5.4   America: Benjamin Franklin

Jumping again to the next century, we find another unlikely fellow involved with magic squares. This person was a printer, publisher, businessman, politician, economist, journalist, scientist, inventor, revolutionary, and diplomat, but he was not a mathematician. He said so himself, saying that he failed math in grade school. You may have guessed the name of this unusual fellow, who was an outstanding American, one of the founding fathers of the United States: Benjamin Franklin (1706–1790). We know about Franklin's kite, his electrical experiments, and his stove, but few people know about old Ben's association with magic squares.

As a young man, Franklin served as clerk to the Assembly, the legislative body of Pennsylvania. In that capacity he had to listen to the politicians' long-winded speeches, but he couldn't take part in them.

It proved to be very boring, and he found that working with magic squares allowed him to escape the boredom. James Logan, a wealthy scholar who had befriended Franklin, had shown Franklin a paper on magic squares by Frénicle. When Logan praised Frénicle's work, Franklin told Logan about his own magic squares that did things beyond those of Frénicle. Logan asked to see Franklin's squares, which he brought on his next visit. Then Logan showed him a paper by Michael Stifel that had a perfect square of order 16, whereas Franklin's square was order 8. Not to be outdone, Franklin went home, and that night made a square of 16 that would outdo Stifel's 16. He showed it to Logan, and, though a modest man, Franklin could not contain himself, declaring "this square of 16 to be the most magically magical of any magic square ever made by any magician" [35, 37, 42].

Logan was greatly impressed by Franklin's squares and wrote about them to a friend in London, Peter Collinson. Years later, when Franklin was in London as representative of the Pennsylvania legislature, Collinson persuaded him to do further work on magic squares, perhaps to make a perfect 16, since Stifel's was a perfect 16 and his was not. Franklin did so and produced a 16 that outshone his own "magically magical" 16. Unfortunately, this new 16 was lost from Franklin's time until just recently when it was discovered in a London museum by Paul Pasles of Georgetown University. We know it to be authentic because it was mentioned in a letter Franklin wrote in 1765 in London to John Canton, a science friend, referring to "the great perfect one of 16" ([39], see section "Patterns Derived by the Present Authors" in chapter 10 for Franklin's perfect 16).

In the same century as Franklin, Leonhard Euler (1707–1783) of Switzerland is credited as one of the most prolific mathematicians who ever lived. He wrote more than 886 papers on all branches of mathematics, not neglecting magic squares. Most amazing is that he accomplished all this while partially or entirely blind for a large part of his life. He devised a new way of making magic squares by following the way the knight moves in a game of chess (Figure 5.9). He also invented Latin squares, a modification of magic squares, that not only gave us the Sudoku puzzle but also has proved to be very useful in agriculture and industry, as will be shown later in the book. In the 1950s, three eminent

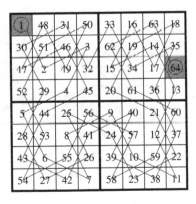

Figure 5.9   The magic line of the knight's tour.

mathematicians working with Latin squares found a conjecture that this acclaimed genius had made, regarding a magic square, was incorrect, and they became known thereafter as Euler's spoilers. Finally, Euler also demonstrated his ability in multiplication by producing a family of 11 children.

5.5   Nineteenth, Twentieth, and Twenty-First Centuries

The nineteenth, twentieth, and twenty-first centuries saw an efflorescence of interest and writing on magic squares. Among the new developments was a relaxation of the rules for making them. For example, from the time of Lo Shu to Benjamin Franklin, squares were constructed using numbers starting with 1 and including the successive numbers, without repetition, to fill the cells in

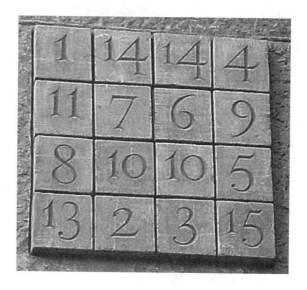

Figure 5.10   Suribach's magic square on facade of
Sagrada Família Cathedral.

Figure 5.11   Gaudí's Sagrada Familia Cathedral, Barcelona.

the square. Now, the use of zero is also permitted. Furthermore, numbers can be repeated and/or skipped. An example of this freer method is the sculptor Suribach's magic square on the wall of the Sagrada Familia Cathedral in Barcelona, Spain (Figures 5.10 and 5.11). This square begins with 1 but there are two 14s and two 10s and in addition, 12 and 16 are missing. This magic square was devised by Suribach because it was part of a facade depicting the death of Christ (Figure 5.12) and Christ died at age 33. Thus the magic constant of this square is 33, rather than 34 as with conventional order 4 magic squares.

Other modern developments in the magic square arena are three-dimensional magic squares, magic cubes, multidimensional squares, even infinite magic squares, geometries other than squares,

Figure 5.12   Suribach's Christ receiving the Judas kiss on
the facade of Sagrada Familia Cathedral.

like magic triangles, magic stars, and so on. In place of simple sums, squares of numbers can be summed and fractions used instead of whole numbers. Thus, there are worlds of possibilities for magic square creators today.

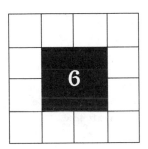

When a Magic Square Is Not Square

6.1 Magic Triangles

The magic triangle shown in Figure 6.1 is presented by Mochalov [105]. It is constructed with the numbers $1, 2, 3, \ldots, 9$.

The magic number is obtained by adding each number along each side of the triangle, that is,

$$8 + 1 + 6 + 5 = 20$$
$$5 + 9 + 4 + 2 = 20 \qquad (6.1)$$
$$2 + 3 + 7 + 8 = 20.$$

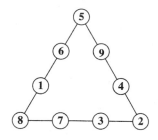

Figure 6.1   Magic triangle.

This magic triangle also has the property that the sum of the squares of each number along each side gives the same number, which is 126, as can be seen below:

$$8^2 + 1^2 + 6^2 + 5^2 = 126$$
$$5^2 + 9^2 + 4^2 + 2^2 = 126 \qquad (6.2)$$
$$2^2 + 3^2 + 7^2 + 8^2 = 126.$$

The four magic triangles shown in Figure 6.2 are constructed with the numbers 1, 2, 3, 4, 5, and 6. Each triangle has a different magic number, as can be verified by adding each number along each side of each triangle.

6.2   Magic Rectangles

When is a square not a square? When it is a rectangle? After all, a rectangle is a square that has grown too much in one dimension. If we can have magic squares, why not magic rectangles?

Definition of Magic Rectangle

A *magic rectangle* is defined as an array of rc integers, where r is the number of rows and c the number of columns, with the following properties (Jelliss [103], Nakamura [106]):

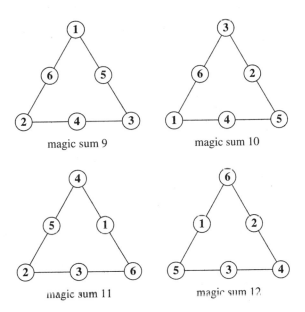

Figure 6.2   Magic triangles with same numbers and
different magic sums.

- the sum of each number in each row gives the same
  value M_r.
- the sum of each number in each column gives the same
  value M_c.

The values M_r and M_c are called the *magic number*, *magic sum*,
or *magic constant* for the magic rectangle. If it refers to M_r it is called
row magic number, and if M_c then it is called *column magic number*.

If the array contains only the integers $1, 2, 3, \ldots, rc$ satisfying
the above properties, then it is called a *standard magic rectangle*. To
obtain the expression for the computation of M_r and M_c, first it is
necessary to find the expression for the sum of the numbers from 1
through rc, which is given by

$$S = \frac{rc}{2}(rc + 1). \tag{6.3}$$

Because the rectangle contains r rows, the row magic number is given by

$$M_r = \frac{S}{r} = \frac{c}{2}(rc + 1). \tag{6.4}$$

The magic number for a rectangle with c columns is obtained from

$$M_c = \frac{S}{c} = \frac{r}{2}(rc + 1). \tag{6.5}$$

Magic Rectangle: Example 1

Figure 6.3(a) shows an $(r \times c) = (even \times even)$ magical rectangle of order $(r \times c) = (2 \times 4)$ with row magic number

$$M_r = \frac{c}{2}(rc + 1) = \frac{4}{2}(2 \times 4 + 1) = 18, \tag{6.6}$$

and column magic number

$$M_c = \frac{r}{2}(rc + 1) = \frac{2}{2}(2 \times 4 + 1) = 9, \tag{6.7}$$

where $r = 2$ rows and $c = 4$ columns.

Magic Rectangle: Example 2

Figure 6.3(b) shows a $(r \times c) = (odd \times odd)$ magical rectangle of order $(r \times c) = (3 \times 7)$ with row magic number

$$M_r = \frac{r}{2}(rc + 1) = \frac{7}{2}(3 \times 7 + 1) = 77, \tag{6.8}$$

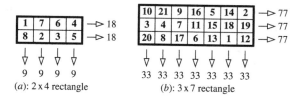

(a): 2 x 4 rectangle

(b): 3 x 7 rectangle

Figure 6.3   (a) An (even × even) magic rectangle;
(b) An (odd × odd) magic rectangle.

and column magic number

$$M_c = \frac{c}{2}(rc + 1) = \frac{3}{2}(3 \times 7 + 1) = 33. \qquad (6.9)$$

Magic Rectangle: Example 3

Thanks to our friend Edward Shineman Jr. of New York, we have a very nice magic rectangle shown in Figure 6.4. He also gave us the magic lightning (Figure 9.8) ([63]).

Figure 6.4 shows a $(r \times c) = (even \times even)$ standard magical rectangle of order $(r \times c) = (16 \times 4)$ with row magic number

$$M_r = \frac{c}{2}(rc + 1) = \frac{16}{2}(4 \times 16 + 1) = 8 \times 65 = 520, \qquad (6.10)$$

and column magic number

$$M_c = \frac{r}{2}(rc + 1) = \frac{4}{2}(4 \times 16 + 1) = 2 \times 65 = 130. \qquad (6.11)$$

Combination of Diagonals

Because each square is perfect, the combination of any four diagonals taken from a total of 8 gives $4 \times 130 = 520$.

Observe, in Figure 6.5, that the main and secondary diagonal of any subsquare make an X, therefore the sum of the numbers in any two Xs will also give the magic sum $4 \times 130 = 520$.

5	32	50	43	21	48	2	59	37	64	18	11	53	16	34	27	→ 520
58	35	13	24	10	51	29	40	26	3	45	56	42	19	61	8	→ 520
47	54	28	1	63	6	44	17	15	22	60	33	31	38	12	49	→ 520
20	9	39	62	36	25	55	14	52	41	7	30	4	57	23	46	→ 520
↓ 130	↓ 130	↓ 130	↓ 130	↓ 130	↓ 130	↓ 130	↓ 130	↓ 130	↓ 130	↓ 130	↓ 130	↓ 130	↓ 130	↓ 130	↓ 130	

Figure 6.4 Magic rectangle of order 16×4.

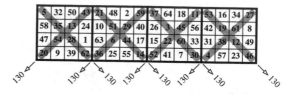

Figure 6.5 Diagonals of each 4 × 4 magic subsquare.

6.3 Magic Cubes

Definition of Magic Cube

Whereas a square is identified by its two dimensions—length, width ($n \times n$)—a cube has three dimensions—length, width, height ($n \times n \times n$). For a magic cube, the dimensions are designated as rows, columns, and pillars. Just as in a magic square, where the number of rows or columns is referred to as the order of a magic square, the order of a magic cube is determined by the rows, columns, or pillars. Likewise, the addition of the numbers in each row, column, or pillar equals the magic constant.

A magic cube is *perfect* if the addition of each number on each row, column, pillar, face diagonals, cross-section diagonals, and space diagonals equals the magic constant [115].

If it is not perfect, then it is called *semi-perfect*.

Example of 3 × 3 × 3 Magic Cube

A magic cube contains the numbers $1, 2, \ldots, n^3$, where n is the order of the magic cube. Therefore, a third-order magic cube ($n = 3$) contains the numbers $1, 2, \ldots, 27$, and its magic constant is given by

$$M_3 = \frac{n}{2}(n^3 + 1) = \frac{3}{2}(3^3 + 1) = \frac{3 \times 28}{2} = 42, \qquad (6.12)$$

where $n = 3$, therefore, $n^3 = 3^3 = 27$.

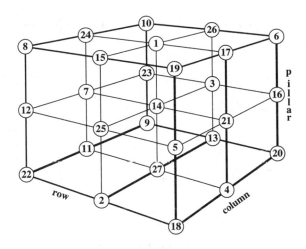

Figure 6.6   Order 3 cube, magic constant of 42.

For example, the sum of each number along each row located on the lower face, shown with thick lines for the order 3 magic cube in Figure 6.6 (Hendricks [27]) equals

$$M_3 = 9 + 11 + 22 = 42$$
$$M_3 = 13 + 27 + 2 = 42$$
$$M_3 = 20 + 4 + 18 = 42. \qquad (6.13)$$

The sum of each number on each column on the lower face, shown with thin lines, equals

$$M_3 = 22 + 2 + 18 = 42$$
$$M_3 = 11 + 27 + 4 = 42$$
$$M_3 = 9 + 13 + 20 = 42. \qquad (6.14)$$

The sum of each number on each pillar, shown with thick lines, on the right face equals

$$M_3 = 18 + 5 + 19 = 42$$
$$M_3 = 4 + 21 + 17 = 42$$
$$M_3 = 20 + 16 + 6 = 42. \qquad (6.15)$$

The sum of each number on each of the four space diagonals equals

$$M_3 = 8 + 14 + 20 = 42$$
$$M_3 = 6 + 14 + 22 = 42$$
$$M_3 = 19 + 14 + 9 = 42$$
$$M_3 = 10 + 14 + 18 = 42. \qquad (6.16)$$

Example of $4 \times 4 \times 4$ Magic Cube

Figure 6.7 shows a fourth-order cube [28], that is, a $4 \times 4 \times 4$ cube. It contains the numbers $1, 2, \ldots, 64$, and its magic number is

$$M_4 = \frac{n}{2}(n^3 + 1) = \frac{4}{2}(4^3 + 1) = \frac{4 \times 65}{2} = 130. \qquad (6.17)$$

For example, each number along each row on the lower face, adds up

$$M_4 = 6 + 51 + 10 + 63 = 130$$
$$M_4 = 27 + 46 + 23 + 34 = 130$$
$$M_4 = 54 + 3 + 58 + 15 = 130$$
$$M_4 = 43 + 30 + 39 + 18 = 130. \qquad (6.18)$$

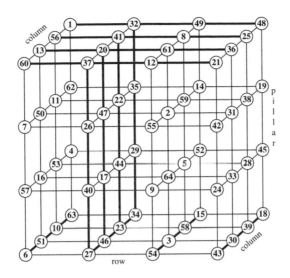

Figure 6.7   A fourth-order cube, magic constant of 130.

The sum of each number along each column on the upper face gives

$$M_4 = 60 + 37 + 12 + 21 = 130$$
$$M_4 = 13 + 20 + 61 + 36 = 130$$
$$M_4 = 56 + 41 + 8 + 25 = 130$$
$$M_4 = 1 + 32 + 49 + 48 = 130. \qquad (6.19)$$

On the right face, the back pillar gives

$$M_4 = 27 + 40 + 26 + 37 = 130$$
$$M_4 = 47 + 17 + 47 + 20 = 130$$
$$M_4 = 23 + 44 + 22 + 41 = 130$$
$$M_4 = 34 + 29 + 35 + 32 = 130. \qquad (6.20)$$

The four diagonals of the magic cube in Figure 6.7 also give the magic sum, namely,

$$M_4 = 6 + 17 + 59 + 48 = 130$$
$$M_4 = 43 + 64 + 22 + 1 = 130$$
$$M_4 = 60 + 47 + 5 + 18 = 130$$
$$M_4 = 21 + 2 + 44 + 63 = 130. \tag{6.21}$$

6.4   Magic Pyramids

A *polyhedron* is a three-dimensional geometric object with flat faces intersecting at segment lines called edges [114].

An *n-sided pyramid* is a polyhedron formed by connecting an n-sided polygonal base and a point, called the apex, by n triangular faces ($n \geq 3$).

A *tetrahedron* is a pyramid with a triangular base, therefore, it has four triangular faces.

A *regular tetrahedron* is a tetrahedron where the four triangular faces are equilateral.

Heinz [93] presents the de Winkel 10-point magic star, which is the intersection of two pyramids, as shown in Figure 6.9. Figure 6.8 shows the two pyramids separated.

Observe that the cross-section shown in gray is the same for both pyramids. Figure 6.9 shows the pyramids in Figure 6.8 intersecting each other and forming the magic pyramids, which contain 22 numbers from 1 through 33, by skipping 11 numbers, 15, 16, 20,..., 26, 28, 31. The magic sum is 36.

The addition of each three numbers along a line gives the same value, which is the magic sum. For example,

$$1 + 33 + 2 = 36, \tag{6.22}$$

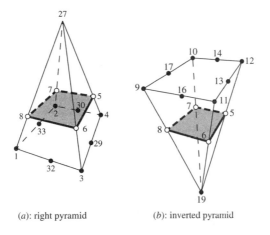

(a): right pyramid   (b): inverted pyramid

Figure 6.8   Pyramids side by side.

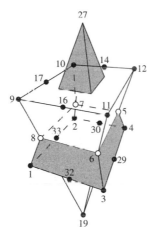

Figure 6.9   Intersection of right and inverted pyramids.

along a line on the base of the standing-up pyramid, see also Figure 6.8(a). And

$$9 + 17 + 10 = 36 \qquad (6.23)$$

along a line on the base of the upside-down pyramid, see also Figure 6.8(b).

The addition of each number along a line connecting one point of the base to the vertex gives

$$1 + 8 + 27 = 36 \qquad (6.24)$$

for the standing-up pyramid, see Figure 6.8(a). And

$$19 + 8 + 9 = 36 \qquad (6.25)$$

for the upside-down pyramid, see Figure 6.8(b).

6.5   Magic Circles

Figures 6.10 and 6.11 show a 4 × 4 magic square with the additional property that the sum of each number along each circle also gives the magic constant, which is 34. Heinz [94] gives credit to Andrews [2] for the idea.

For example, from Figure 6.10, along the small circle on the upper left,

$$M_4 = 1 + 6 + 16 + 11 = 34, \qquad (6.26)$$

and along the dotted circle

$$M_4 = 1 + 15 + 4 + 14 = 34. \qquad (6.27)$$

Figure 6.11 shows a different arrangement of circles where the addition of each number along each circle also gives the magic constant.

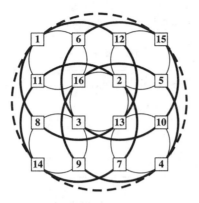

Figure 6.10   Intersection of magic circles and magic square.

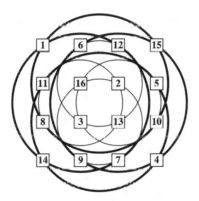

Figure 6.11   Intersection of magic circles and magic square.

The three examples that follow illustrate the computation of the magic number along each circle.

The sum of each number in Figure 6.11 along the dark leftmost circle gives

$$M_4 = 1 + 12 + 7 + 14 = 34. \qquad (6.28)$$

Along the small, lighter lower circle the sum is

$$M_4 = 16 + 2 + 7 + 9 = 34. \tag{6.29}$$

At the center of the smaller, lighter circle each number adds

$$M_4 = 16 + 2 + 13 + 3 = 34. \tag{6.30}$$

6.6   Magic Spheres

It makes no sense to say *row* or *column* on a sphere, so we say *parallel* and *meridian*. The magic sum for the sphere in Figure 6.12 can be evaluated from a meridian as

$$14 + 10 + 24 + 21 + 13 + 17 + 3 + 6 = 180, \tag{6.31}$$

or along a parallel as

$$1 + 24 + 2 + 23 + 26 + 3 + 25 + 4 = 108. \tag{6.32}$$

Although the parallels could be paired with the rows, the meridians have no relation to the columns. Observe

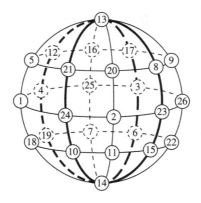

Figure 6.12   Magic sphere.

Figure 6.12: the four meridians intersect at the north pole, where the cell contains the number 13, and also at the south pole over the cell with the number 14. Therefore, these numbers belong to four meridians simultaneously.

6.7   Magic Stars

Standard and Nonstandard Magic Stars

A magic star is called *standard* if it contains each of the integers $1, 2, 3, \ldots, 2n$, where n is the number of points of the star. Otherwise, it is called *nonstandard*. See Figure 6.14 for examples of standard magic stars and Figure 6.15 for an example of a nonstandard magic star.

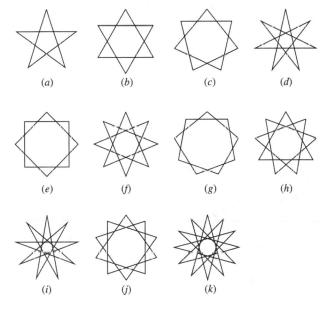

(a)   (b)   (c)   (d)

(e)   (f)   (g)   (h)

(i)   (j)   (k)

Figure 6.13   Types of stars.

Magic Number of a Star

The star will be magic if the addition of each number on each chord line gives the same value, which is called the *star magic number*. The chord lines are illustrated in Figures 6.14 and 6.15. For a standard star it is given by

$$M_{star} = 4n + 2, \qquad (6.33)$$

where n is the number of the points of the star. For the derivation of the expression for the star magic number in (6.33) see "Evaluation of the magic number of a star" in chapter 15.

Examples of Standard Magic Stars

Figure 6.14(a) shows a magic hexagram, that is, a star constructed over a hexagon. It uses the numbers $1, 2, \ldots, 12$, and the magic number is given by $M_6 = 4n + 2 = 4 \times 6 + 2 = 26$.

Figure 6.14(b) shows a star constructed over a heptagon, that is, a heptagram, and the magic number is $M_7 = 4n+2 = 4 \times 7+2 = 30$. This magic star uses the numbers $1, 2, \ldots, 2n$, which for $n = 7$ gives $1, 2, \ldots, 14$.

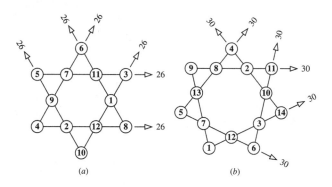

(a)   (b)

Figure 6.14 (a) Magic hexagram; (b) Magic heptagram.

Examples of Nonstandard Magic Stars

Example 1: Nonstandard Pentagram

The magic star shown in Figure 6.15 is constructed over a pentagram 6.13(a), which is the smallest possible star, but it is not a standard magic star (Heinz [91]). Instead of having the numbers $1, 2, \ldots, 10$ as it would be for a standard $n = 5$ magic star, it contains the numbers $1, 2, \ldots, 12$, skipping 7 and 11.

Magic Stars of Type S and T

Trenkler [108] classifies the magic stars with four numbers in each chord line as

- type S if all numbers are along the contour of the star. For example, the stars in Figures 6.14(a) and (b) are of type S.
- type T if the numbers are in each point and inside the contour of the star, as shown in Figure 6.16.

The magic star in Figure 6.16 is a standard magic star with $n = 8$ points, Figure 6.13(f), constructed with the numbers $\{1, 2, 3, \ldots, 16\}$, where $2n = 16$. The magic constant is given by $M_8 = 4n + 2 = 4 \times 8 + 2 = 34$.

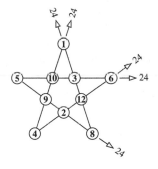

Figure 6.15   This pentagram is a nonstandard magic star.

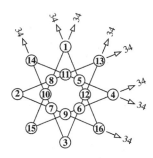

Figure 6.16   Type T octagram.

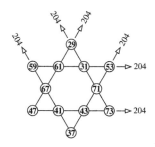

Figure 6.17   A magic star with consecutive prime numbers.

Magic Star with Prime Numbers

Trenkler [108] also exhibits a hexagram constructed with the following consecutive prime numbers: 29, 31, 37, 41, 43, 47, 53, 59, 61, 67, 71, 73, which is shown in Figure 6.17 and has a magic constant of 204. It is a nonstandard magic star of type S with six points as shown in Figure 6.13(b).

Magic Star Exercise for Schoolchildren in Adding Fractions

The hexagram shown in Figure 6.18 was found by Heinz [92] in a "grade five exercise booklet" by Larson [41].

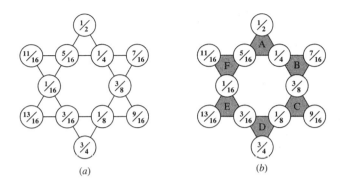

(a)   (b)

Figure 6.18   (a) Magic hexagram with fractions—star magic constant is 27/16; (b) Small triangles in the hexagram.

The exercise, based on Figure 6.18, contains the following questions:

1. Add the four fractions in each of the six lines.
2. Are the six sums the same?
3. Add the three fractions in each of the two big triangles.
4. Are the two sums the same?
5. Add the three fractions in each of the six small triangles.
6. Are the six sums the same?

Solution to the Exercise

- Item 1 is asking to evaluate the sum of each fraction along each chord line. There are six chord lines with four fractions on each. For example, the chord lines in the gray triangle in Figure 6.19 gives

$$\frac{1}{2} + \frac{1}{4} + \frac{3}{8} + \frac{9}{16} = \frac{27}{16}$$

$$\frac{1}{2} + \frac{5}{16} + \frac{1}{16} + \frac{13}{16} = \frac{27}{16}$$

$$\frac{13}{16} + \frac{3}{16} + \frac{1}{8} + \frac{9}{16} = \frac{27}{16}$$

$$\frac{11}{16} + \frac{5}{16} + \frac{1}{4} + \frac{7}{16} = \frac{27}{16}$$

$$\frac{11}{16} + \frac{1}{16} + \frac{3}{16} + \frac{3}{4} = \frac{27}{16}$$

$$\frac{7}{16} + \frac{3}{8} + \frac{1}{8} + \frac{3}{4} = \frac{27}{16} \qquad (6.34)$$

- Item 2 asks if the six sums are the same, which is the case. Therefore, each sum equals the star magic number, as can be seen from (6.34).
- Item 3 asks to add three fractions in each of the two big triangles. Figure 6.19(a) and (b) shows the two big triangles. Adding the three fractions from the corners of the big triangle shown in gray in Figure 6.19(a) gives

$$\frac{1}{2} + \frac{9}{16} + \frac{13}{16} = \frac{30}{16}. \qquad (6.35)$$

Adding the three fractions from the corners of the big triangle shown in gray in Figure 6.19(b) it gives

$$\frac{11}{16} + \frac{5}{16} + \frac{3}{8} = \frac{22}{16}. \qquad (6.36)$$

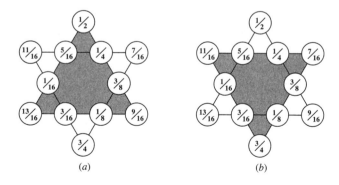

(a)   (b)

Figure 6.19   Big triangles in the hexagram.

- Item 4 asks: Are the two sums the same? They are not the same, but it is an addition of fractions exercise.
- Items 5 and 6 ask to add the three fractions in the small triangles shown in gray in Figure 6.18(b). The sum of the numbers in each small gray triangle equals 17/16, for the gray triangle B it gives

$$\frac{1}{4} + \frac{7}{16} + \frac{3}{8} = \frac{17}{16}, \qquad (6.37)$$

and for the gray triangle E it yields

$$\frac{1}{16} + \frac{3}{16} + \frac{13}{16} = \frac{17}{16}. \qquad (6.38)$$

The hexagram in Figure 6.18 or 6.19 contains the numbers $1/16, 2/16, \ldots, 13/16$, skipping $10/16$. The star magic constant is $27/16$.

Probably Larson constructed a magic hexagram with the numbers $1, 2, \ldots, 9, 11, 12, 13$, skipping the 10 and 14 to be able to assign numbers to the small triangles such that the sum of each number would give the same value. Then he divided each number by 16 to transform the hexagram into an addition of fractions exercise.

6.8   Magic Pentagrams

This beauty by L. S. Frierson [3; Figure 6.20], created in the nineteenth century, is a magic pentagram (a star with five points) made up of five diamonds. This star has many interesting properties that are displayed in Figures 6.21–6.24. This one magic pentagram offers many different magic combinations within one figure. Each 4×4 diamond is perfectly magic with a magic sum of 162, as shown in Figure 6.21 for the rows and columns and Figure 6.22 for the diagonals. Each diamond contains 2×2 diamonds, shown in gray in Figures 6.21 and 6.23, that have a sum of each number

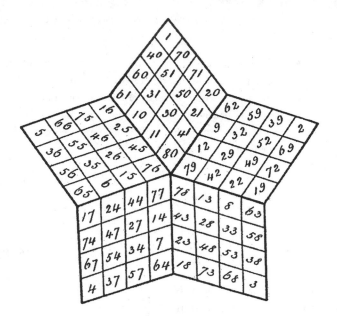

Figure 6.20 Magic pentagram star.

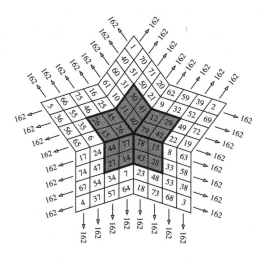

Figure 6.21 Rows, columns, and 2 × 2 diamonds around the center. The sum of each number on each row and column of each diamond equals the magic constant 162.

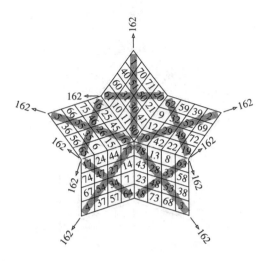

Figure 6.22 The sum of each number on each diagonal of each diamond equals the magic constant 162.

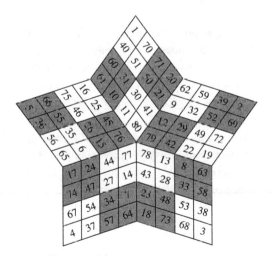

Figure 6.23 The sum of each number on each 2 × 2 diamond gives the magic constant of 162.

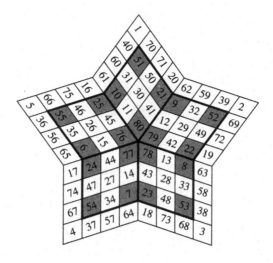

Figure 6.24 Four corners on each 3 × 3 diamond, shown in gray, around the center. The sum of each number on each corner of each diamond equals the magic constant 162.

equaling the magic sum 162. Each 4 × 4 diamond also contains four 3 × 3 diamonds whose corner numbers add up to 162, thus 76 + 6 + 55 + 25 = 162 (Figure 6.24). The tips of the pentagram contain the numbers 1, 2, 3, 4, 5, and its center numbers are also consecutive, 76, 77, 78, 79, 80. Note that all the numbers in the top diamond end in 0 or 1, for example 1, 70, 71, 20; in the next diamond going clockwise they end in 2 or 9, for example 62, 59, 59, 2; in the next diamond, 3 or 8, for example 78, 13, 8, 63; in the fourth diamond, 4 or 7, for example, 17, 24, 44, 77; and in the fifth diamond, 5 or 6, for example 16, 25, 45, 76.

6.9   Magic Crosses

The magic cross (Figure 6.25) is also the ingenious creation of nineteenth-century magic man L. S. Frierson [4].

Figure 6.25 Magic cross.

Description of the Cross

The magic cross is made up of four rectangles called top, bottom, left, and right. The left and right rectangles are also called the arms of the cross.

The top of the cross has four columns and eight rows, the bottom has four columns and twelve rows, while each left and right arm has four rows and eight columns. The center of the cross contains the number 1. The rows and columns in each rectangle of the cross are numbered from the top left corner.

The magic constant 1,471 is obtained by the addition of 21 numbers. For example, starting on the left column on top, moving to the number 1, and then to the second column from the left on the bottom, the 21 numbers write:

$$2 + 91 + 128 + 73 + 3 + 90 + 129 + 72$$
$$+ 1 +$$
$$62 + 139 + 80 + 13 + 63 + 138 + 81 +$$
$$12 + 64 + 137 + 82 + 11 = 1,471. \qquad (6.39)$$

Or, starting on the first row on the left arm, moving to the number 1 and then to the second column from the left on the bottom, the 21 numbers write:

$$4 + 58 + 125 + 107 + 5 + 59 + 124 + 106$$
$$+ 1 +$$
$$62 + 139 + 80 + 13 + 63 + 138 + 81 +$$
$$12 + 64 + 137 + 82 + 11 = 1,471. \qquad (6.40)$$

It is said that the cross contains the almost incredible number of 160,144 combinations of rows and columns of 21 numbers that give the magic constant of 1,471. Now that you've gone this far, you have only 160,142 to find!

The reader may wish to work on his or her own with the magic cross to obtain 21 numbers such that their sum gives the magic constant 1,471. But if you want to save time, you might be better advised to read the text that follows.

Computation of the Magic Constant

The following can be identified from Figure 6.26:

- The top and bottom of the cross have four columns each, therefore each of their rows contains four numbers.

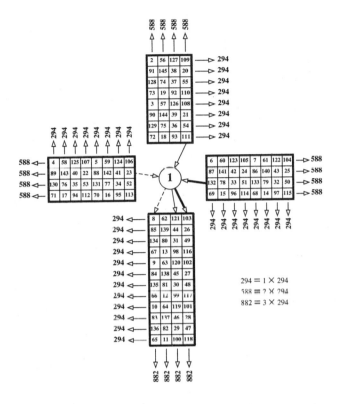

Figure 6.26 Three possible paths for obtaining the magic number.

The sum of each number along each row on the top or bottom of the cross gives 294.

- The two arms have four rows each, therefore each of their columns have four numbers. The sum of the four numbers on each column gives 294.
- The sum of each number along each column on the top equals 2 × 294 = 588.
- The sum of each number along each column on the bottom of the cross equals 3 × 294 = 882.
- The sum of each number along any column on the left plus the sum of any column on the right arm gives 2 × 294 = 588.

The magic constant is defined as the sum of 20 numbers plus the number 1, which is the value on the center of the cross. As a consequence of the choice of selection of the numbers on the top, bottom, and two arms such that the sum of the numbers on each row or column gives a multiple of 294, the magic constant of the cross can be written, in terms of 294, as $M_{cross} = 5 \times 294 = 1,470$.

Example of paths for the computation of the magic constant.

Example 1. Path: fourth column of the upper arm, move to the 1 in the center along the arrow with the thin solid line, then to the third column of the lower arm as indicated in Figure 6.26. The numbers along this path are

$$M_{cross} = (109 + 20 + 55 + 110 + 108 + 21 + 54 + 111)$$
$$+ 1 +$$
$$(121 + 44 + 31 + 98 + 120 + 45 + 30$$
$$+ 99 + 119 + 46 + 29 + 100)$$
$$= 588 + 1 + 882 = 1,471. \tag{6.41}$$

Example 2. Path: second row of the left arm, move to the center along the arrow with dashed line, then to the first column of the lower arm as indicated in Figure 6.26. The numbers along this path are

$$M_{cross} = (89 + 143 + 40 + 22 + 88 + 142 + 41 + 23)$$
$$+ 1 +$$
$$(8 + 85 + 134 + 67 + 9 + 84 + 135$$
$$+ 66 + 10 + 33 + 136 + 65)$$
$$= 588 + 1 + 882 = 1,471. \tag{6.42}$$

Example 3. Path: third row of the right arm, move to the center along the arrow with the thick line, then to the fourth column of the lower arm as indicated in Figure 6.26. The numbers along this path are

$$M_{cross} = (50 + 32 + 79 + 133 + 51 + 33 + 78 + 132)$$
$$+ 1 +$$
$$(103 + 26 + 49 + 116 + 102 + 27 + 48 + 117$$
$$+ 101 + 28 + 47 + 118)$$
$$= 588 + 1 + 882 = 1{,}471. \tag{6.43}$$

How Many Paths Are There to Obtain the Cross Magic Constant?

The number of possible paths using this approach can be obtained by observing that each column in the top rectangle, for example, can go to four different columns in the bottom rectangle. Therefore, it gives a total of $4 \times 4 = 16$ different paths. The same is true if the paths starts on the left or right arm. This gives a total of $3 \times 16 = 48$ different paths.

Any path that sums to $5 \times 294 = 1{,}470$ will be considered as a valid path. For example

- Figure 6.27 shows one of the many possible choices for the computation of the magic constant. The arrows in Figure 6.27 indicate that
  - any column from the upper arm (2×294), plus
  - any column from the left arm (1×294), plus
  - any column from the right arm (1×294), plus
  - any row from the lower arm (1×294),
  gives the magic constant, that is

$$(2 \times 294) + 3 \times (1 \times 294) = 5 \times 294 = 1{,}471 \tag{6.44}$$

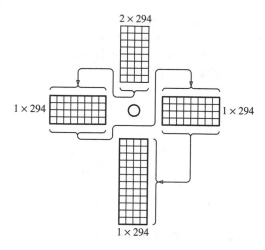

Figure 6.27   Possible paths to obtain the magic constant.

The number of choices for the evaluation of the magic constant using the path shown in Figure 6.27 can be obtained as follows:

- Contribution of the top rectangle: The top rectangle has four columns, therefore, there are four different cases for the computation of the magic constant.
- Contribution of the left arm: The left arm is a rectangle with eight columns to be chosen from. The association of each column from the top rectangle with each column from the left arm gives $4 \times 8 = 32$ different choices.
- Contribution of the right arm: The right arm is a rectangle with eight columns to be chosen from. Each of the 32 previous choices can be selected to choose any one of the eight columns of the right arm, which gives an accumulated number of choices equal to $32 \times 8 = 256$.

- Contribution of the bottom rectangle: The bottom rectangle has 12 rows to be chosen from. Each of the 256 previous choices associated to each of the 12 rows gives $256 \times 12 = 3{,}072$ different choices for the evaluation of the magic constant of the cross.

  In summary the number of choices is

$$4 \times 8 \times 8 \times 32 = 3{,}072. \qquad (6.45)$$

- If, in the above case, instead of moving to the left arm first, move to the right and then to the left and then to the bottom, there will be another 3,072 possible cases.

- One can also choose to start from one of the four rows on the left arm (2×294), plus one of the eight rows of the top rectangle (1×294), plus one of the eight columns of the right arm (1×294), plus one of the 12 rows of the bottom rectangle (1×294). This gives a total of $4 \times 8 \times 8 \times 12 = 3{,}072$ possible cases.

- Each column of the bottom rectangle (3×294), one column of the right arm, and one column of the left arm, which gives $4 \times 8 \times 8 = 256$ possible cases.

- Any column from the lower arm (3×294) plus any column from both side arms (2×294) give the magic constant, that is

$$3 \times 294 + 2 \times 294 = 5 \times 294 = 1{,}471. \qquad (6.46)$$

As an exercise, try to find more paths that give the magic constant. Pickover [57] says that there are 160,144 different tracks of 21 numbers that give the magic constant.

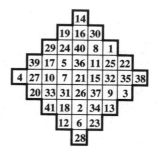

Figure 6.28   Serrated magic square.

6.10   Magic Serrated Edge Figures

The serrated magic square (see Figure 6.28) can be seen as a magic square that the mice chewed on. The serrated magic square can be found in [5], [20], and [60].

Properties of the Serrated Magic Square

Figure 6.29 shows a pattern composed of two parallel rows that add up to the magic constant of 189. There are two more similar patterns, which are left to the reader to find. Figure 6.30(a) shows the row pattern represented in the serrated square, and Figure 6.30(b) shows the same row pattern represented in a square.

Any combination of the numbers adding to 105 with the numbers adding to 84 gives the magic number of 189. It is left to the reader to find out how many different combinations can be found adding up to give the magic constant.

For example, the numbers in the gray area in Figure 6.30 add up:

$$(20 + 10 + 5 + 40 + 30) + (33 + 7 + 36 + 8)$$
$$= 105 + 84 = 189. \tag{6.47}$$

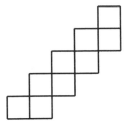

Figure 6.29 Representation of the rows in a serrated magic square.

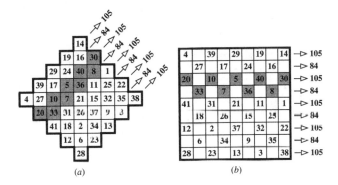

Figure 6.30 Representation of the row pattern.

Figure 6.31 shows a pattern composed of two parallel columns that add up the magic number to 189. There are two more similar patterns, which are left to the reader to find. Figure 6.32(a) shows the columns pattern represented in the serrated square, and Figure 6.32(b) shows the same row pattern represented in a square.

For example, the numbers in the gray area in Figure 6.32 add up like so:

$$(39 + 10 + 31 + 2 + 23) + (17 + 7 + 26 + 34) = 105 + 84 = 189. \tag{6.48}$$

The main diagonal numbers in Figure 6.33 add up:

$$4 + 27 + 10 + 7 + 21 + 15 + 32 + 35 + 38 = 189. \tag{6.49}$$

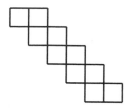

Figure 6.31 Representation of the columns pattern.

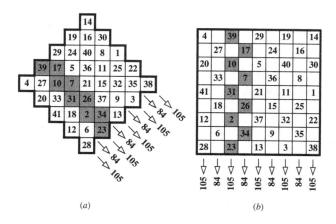

(a)

(b)

Figure 6.32 Representation of the columns.

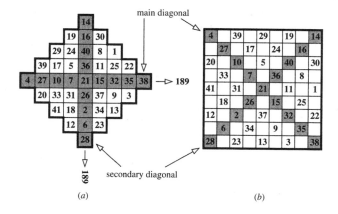

(a)

(b)

Figure 6.33 Representation of the diagonals.

Figure 6.34   Order 3 square pattern.

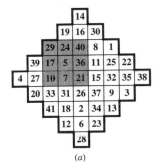

4		39		29		19		14
	27		17		24		16	
20		10		5		40		30
	33		7		36		8	
41		31		21		11		1
	18		26		15		25	
12		2		37		32		22
	6		34		9		35	
28		23		13		3		38

(a)   (b)

Figure 6.35   Subsquare of order 3 adding up to the magic number.

And the secondary diagonal numbers in Figure 6.33 add up:

$$14 + 16 + 40 + 36 + 21 + 26 + 2 + 6 + 28 = 189. \quad (6.50)$$

Figure 6.34 shows an order 3 square used as a pattern. The addition of all numbers in some 3 × 3 subsquares gives the magic number 189.

Figure 6.35(a) shows a case using the pattern in the serrated square, which adds up to the magic number 189. Figure 6.35(b) shows the same pattern represented in a square.

The numbers in the 3 × 3 subsquare in gray add up:

$$29 + 24 + 40 + 17 + 5 + 36 + 10 + 7 + 21 = 189. \quad (6.51)$$

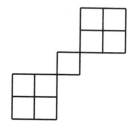

Figure 6.36 Pattern in the row direction adding up to the magic constant.

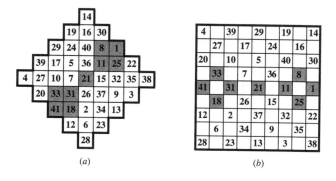

(a)   (b)

Figure 6.37 Pattern adding up to the magic constant.

The gray square in Figure 6.35 can move from left to right and top to bottom and still gives the magic number.

Figure 6.36 shows a pattern parallel to the rows, which adds up to the magic number 189. There are two more similar patterns, which are left to the reader to find. Figure 6.37(a) shows the pattern represented in the serrated square and Figure 6.37(b) shows the same pattern represented in a square.

The numbers in the pattern in gray in Figure 6.37 add up:

$$(33 + 31 + 41 + 18) + 21 + (8 + 1 + 11 + 25) = 189. \quad (6.52)$$

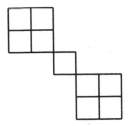

Figure 6.38 Pattern adding up to the magic number.

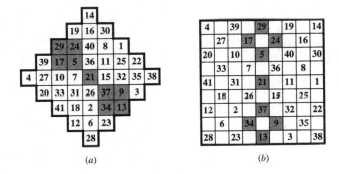

Figure 6.39 Pattern in the column direction adding up
to the magic constant.

Figure 6.38 shows a pattern parallel to the columns, which add up to the magic constant of 189. There are two more similar patterns, which are left to the reader to find. Figure 6.39(a) shows the pattern represented in the serrated square and Figure 6.39(b) shows the same pattern represented in a square.

The numbers in the pattern in gray in Figure 6.39 add up:

$$(29 + 24 + 17 + 5) + 21 + (37 + 9 + 34 + 13) = 189. \quad (6.53)$$

Figure 6.40 shows a cross pattern made of five cells along each arm. Only some of the crosses add up to the magic constant of 189.

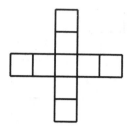

Figure 6.40   Cross pattern.

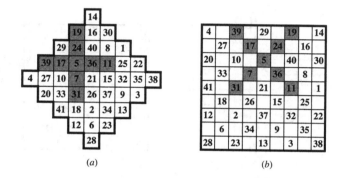

(a)   (b)

Figure 6.41   Cross pattern adding up to the magic number.

Figure 6.41(a) shows the cross pattern represented in the serrated square, and Figure 6.41(b) shows the same pattern represented in a square.

The numbers in the cross pattern in gray in Figure 6.41 add up:

$$(19 + 24 + 7 + 31) + 5 + (39 + 17 + 36 + 11) = 189. \quad (6.54)$$

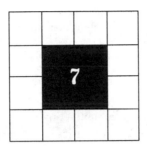

Magic Squares and Arithmetic

The unexpected coincidence of mathematical regularities that produces the mystery of magic squares by addition carries over to other mathematical calculations as well. It engenders surprise and pleasure in the squares of the magic curiosities that will follow.

7.1 Magic Square Subtraction

This and the next two arithmetical calculations, multiplication and division, come this way courtesy of math gamester Dudeny [14].

Figure 7.1   Subtraction magic square.

In Figure 7.1, if the middle number is subtracted from the sum of the two end numbers, that result is the magic constant 5. For the top row, for instance, $2 + 4 = 6$ and $6 - 1 = 5$. You may achieve the same result by subtracting the first number in a line from the second $1 - 2$ and that result subtracted from the third $4 - (1 - 2) = 4 - (-1) = 4 + 1 = 5$, or by going in the opposite direction $2 - (1 - 4) = 2 - (-3) = 2 + 3 = 5$. However, to avoid negative numbers the first method is preferred. Note that the diagonals also give the magic sum, that is, $4 - (5 - 6) = 5$ and $2 - (5 - 8) = 5$.

7.2   Magic Square Multiplication

In Figure 7.2, the magic constant 216 is obtained by multiplying the numbers in any line. Thus, for the main diagonal, $3 \times 6 \times 12 = 18 \times 12 = 216$ and for the secondary diagonal, $18 \times 6 \times 2 = 18 \times 12 = 216$. The product of each number in first row give $12 \times 1 \times 18 = 12 \times 18 = 216$. See Dudeny [14].

7.3   Magic Square Division

In the division square, Figure 7.3, the magic constant 6 is obtained by dividing the second number in a line by the first, in either direction, and the third number by the quotient.

12	1	18
9	6	4
2	36	3

Figure 7.2   Multiplication magic square.

3	1	2
9	6	4
18	36	12

Figure 7.3   Division magic square.

Therefore, to compute the value for each row the following operations must be performed:

third column ÷ (second column ÷ first column),   **(7.1)**

exemplifying, for the first row,

$$2 \div (1 \div 3) = 2 \div \frac{1}{3} = 2 \times \frac{3}{1} = 6;$$ **(7.2)**

for the second row,

$$4 \div (6 \div 9) = 4 \div \frac{6}{9} = 4 \times \frac{9}{6} = 6.$$ **(7.3)**

A similar procedure applies for the computation related to each column, that is,

$$\text{third row} \div (\text{second row} \div \text{first row}). \qquad (7.4)$$

For the first column, for example, it gives

$$18 \div (9 \div 3) = 18 \div \frac{9}{3} = 18 \times \frac{3}{9} = 6. \qquad (7.5)$$

For the main diagonal, the operations are

$$12 \div (6 \div 3) = 12 \div \frac{6}{3} = 12 \times \frac{3}{6} = 6. \qquad (7.6)$$

To avoid fractions, you can divide the product of the two end numbers by the middle number. Thus, for the diagonal, $3 \times 12 = 36$ and 36 divided by 6 equals 6, that is $38 \div 6 = 6$. See Dudeny [14].

7.4   Addition-Multiplication Magic Square

Walter W. Horner [31], retired math teacher of Pittsburgh, has given us an unusual square that is simultaneously magic in both addition and multiplication (Figure 7.4). It has a magic constant of 840 for the addition of each number in each row, column, or diagonal. It also has a magic constant of 2,058,068,231,856,000 for the multiplication of each number in each row, column, or diagonal.

7.5   Bimagic Square

Like the addition-multiplication square, the bimagic square is doubly magic. It is magic both in addition and again when each of its numbers is squared and then added. This order 9 square, shown in Figure 7.5, was designed by master magic square creator

46	81	117	102	15	76	200	203
19	60	232	175	54	69	153	78
216	161	17	52	171	90	58	75
135	114	50	87	184	189	13	68
150	261	45	38	91	136	92	27
119	104	108	23	174	225	57	30
116	25	133	120	51	26	162	207
39	34	138	243	100	29	105	152

Figure 7.4 Addition-multiplication magic square.

1	23	18	33	52	38	62	75	67
48	40	35	77	72	55	25	11	6
65	60	79	13	8	21	45	28	50
43	29	51	66	58	80	14	9	19
63	73	68	2	24	16	31	53	39
26	12	4	46	41	36	78	70	56
76	71	57	27	10	5	47	42	34
15	7	20	44	30	49	64	59	81
32	54	37	61	74	69	3	22	17

Figure 7.5 Bimagic square.

1	22	33	41	62	66	79	83	104	112	123	144
9	119	45	115	107	93	52	38	30	100	26	136
75	141	35	48	57	14	131	88	97	110	4	70
74	8	106	49	12	43	102	133	96	39	137	71
140	101	124	42	60	37	108	85	103	21	44	5
122	76	142	86	67	126	19	78	59	3	69	23
55	27	95	135	130	89	56	15	10	50	118	90
132	117	68	91	11	99	46	134	54	77	28	13
73	64	2	121	109	32	113	36	24	143	81	72
58	98	84	116	138	16	129	7	29	61	47	87
80	34	105	6	92	127	18	53	139	40	111	65
51	63	31	20	25	128	17	120	125	114	82	94

Figure 7.6   Order 12 perfect trimagic square of Walter Trump.

John Hendricks of Canada in 1999 (Heinz and Hendricks [19]). With addition, the magic sum is 369, and when the numbers are squared the magic sum is 20,049. Furthermore, when all the numbers in each 3 × 3 square are added, they give the magic sum of 369. The smallest bimagic square is of order 8 and the smallest trimagic square, that is, a bimagic square that is also magic in the sum when the numbers are cubed, is of order 12, as in Figure 7.6. A trimagic square need not be bimagic as well.

A square is said to be P-magic if it is still magic after the numbers in each cell have been replaced by their k-th power (for $k = 1, 2, \ldots, P$). For example, the square in Figure 7.5 is bimagic (2-magic) and in Figure 7.6 is trimagic (3-magic).

The sum of the numbers in each cell along any row is 870. For example, for the rows 2 and 8, it yields

$$9 + 119 + 45 + 115 + 107 + 93 + 52 + 38 + 30$$
$$+ 100 + 26 + 136 = 870$$

$$132 + 117 + 68 + 91 + 11 + 99 + 46 + 134 + 54$$
$$+ 77 + 28 + 13 = 870. \tag{7.7}$$

The sum of the squares of the numbers in each cell along any row is 83,810. For the same rows above it is

$$9^2 + 119^2 + 45^2 + 115^2 + 107^2 + 93^2 + 52^2 + 38^2 + 30^2$$
$$+ 100^2 + 26^2 + 136^2 = 83,810$$
$$132^2 + 117^2 + 68^2 + 91^2 + 11^2 + 99^2 + 46^2 + 134^2 + 54^2$$
$$+ 77^2 + 28^2 + 13^2 = 83,810. \tag{7.8}$$

The sum of the cubes of the numbers in each cell along any row is 9,082,800. For the same rows above it is

$$9^3 + 119^3 + 45^3 + 115^3 + 107^3 + 93^3 + 52^3 + 38^3 + 30^3$$
$$+ 100^3 + 26^3 + 136^3 = 9,082,800$$
$$132^3 + 117^3 + 68^3 + 91^3 + 11^3 + 99^3 + 46^3 + 134^3 + 54^3$$
$$+ 77^3 + 28^3 + 13^3 = 9,082,800. \tag{7.9}$$

Similarly the columns and diagonals give 870 for the sum of each number, 83,810 for the sum of the square of each number, and 9,082,800 for the sum of the cubic of each number.

The magic constant for each case is evaluated by the following mathematics:

1. Sum of each number:

$$M_{12} = \frac{n}{2}\left(n^2 + 1\right) = \frac{12}{2}\left(12^2 + 1\right) = 870. \tag{7.10}$$

2. Sum of the square of each number.

$$
M_{12}^2 = \frac{1}{n} \left[\frac{n^2 \left(n^2 + 1\right) \left(2n^2 + 1\right)}{2 \times 3} \right]
$$

$$
= \frac{1}{12} \left[\frac{12^2 \left(12^2 + 1\right) \left(2 \times 12^2 + 1\right)}{2 \times 3} \right] = 83,810.
$$

(7.11)

The formula for computing the sum of the squares of the numbers $1, 2, \ldots, 144$, is found in Tavares [73].

3. Sum of the cubes of each number.

$$
M_{12}^3 = \frac{n^4}{4} \left(n^2 + 1\right)^2
$$

$$
= \frac{12^4}{4} \left(12^2 + 1\right)^2 = 9,082,800.
$$

(7.12)

The expression for the computation of the sum of the cubes of the numbers $1, 2, \ldots, 144$, is in Tavares [73].

7.6   Geometric Magic Square

Definition and Construction

A geometric magic square is made up of a base number (in Figure 7.7 the base number is 2) raised to the powers of the exponents (in this case from 1 to 9) to give another square. For the example in Figure 7.7, the product of each line yields the magic constant of 32,768. This beauty is displayed for us by Donald D. Spencer ([65]).

To construct a geometric magic square, choose any value among the positive integers to be the base and any addition magic square. Use the numbers of the magic square as exponents for the chosen base. In Figure 7.7, the chosen base is 2 and the magic square is a 3 × 3 perfect square, shown in Figure 7.8.

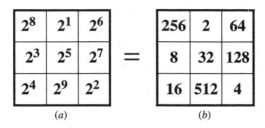

2^8	2^1	2^6
2^3	2^5	2^7
2^4	2^9	2^2

$=$

256	2	64
8	32	128
16	512	4

(a)   (b)

Figure 7.7 Geometric magic square.

8	1	6
3	5	7
4	9	2

Figure 7.8 Addition magic square, whose numbers are used as exponents.

As a consequence of using the numbers in an addition as exponents, the sum of each exponent in each row, column, or diagonal equals 15, the magic sum for a perfect 3 × 3 magic square like the one shown in Figure 7.8. For example, the first row gives

$$2^8 \times 2^1 \times 2^6 = 2^{8+1+6} = 2^{15} = 32,768. \qquad (7.13)$$

Therefore, the product of the numbers in each row, column, and diagonal will always give the same number.

Benjamin Franklin Geometric Magic Square

Figure 7.10 shows a 4 × 4 geometric magic square based on the Benjamin Franklin perfect 4 × 4 magic square shown in Figure 7.9.

16	3	2	13
5	10	11	8
9	6	7	12
4	15	14	1

Figure 7.9   Benjamin Franklin order 4 magic square.

3^{16}	3^3	3^2	3^{13}		43046721	27	9	1594323
3^5	3^{10}	3^{11}	3^8	$=$	243	59049	177147	6561
3^9	3^6	3^7	3^{12}		19683	729	2187	531441
3^4	3^{15}	3^{14}	3^1		81	14348907	4782969	3

(a) (b)

Figure 7.10   Geometric magic square based on the Benjamin
Franklin order 4 magic square.

The magic product is $3^{34} = 16, 677, 181, 699, 666, 569$, which can be approximated to $1.66771817 \times 10^{16}$.

Note that Figure 7.10(b) is constructed as a rectangle rather than as a square. This procedure is followed in other cases in this book where large numbers are produced.

Benjamin Franklin's "Four Plus Five Equals Nine" Magic Square

The magic square in Figure 7.11 is a 5 × 5 with the Benjamin Franklin 4 × 4 embedded in it. It was constructed using a technique shown by Heinz at his Web site [95].

The 5 × 5 magic square contains the numbers 1, 2, . . . , 25 and a magic sum 65.

The 4 × 4 magic square, whose numbers are in the small squares, was obtained from the Benjamin Franklin 4 × 4 magic square, Figure 7.9, by adding 25 to each number. The addition of a constant value to each number does not change the magic properties, and the new magic sum is $34 + 25 × 4 = 134$.

The magic constant for the 9 × 9 magic square is obtained by adding each number along each row of the 5 × 5 and along each row composed of the small squares. The magic constant equals the addition of the magic constant of both magic squares, that is, $65 + 134 = 199$.

For example, the addition of each number along the first row gives

$$15 + 41 + 18 + 28 + 1 + 27 + 24 + 38 + 17 = 199, \quad (7.14)$$

or, identifying, between parentheses, the numbers belonging to each square in Figure 7.11, equation (7.14) rewrites

$$(15 + 18 + 1 + 24 + 17) + (41 + 28 + 27 + 38)$$
$$= 65 + 134 = 199. \quad (7.15)$$

There are two possibilities to obtain addition of each number along the second row. First case

$$16 + 41 + 14 + 28 + 7 + 27 + 5 + 38 + 23 = 199. \quad (7.16)$$

Second case

$$16 + 30 + 14 + 35 + 7 + 36 + 5 + 33 + 23 = 199. \quad (7.17)$$

The properties shown in equations (7.16) and (7.17) are valid for the addition of each number in any row or column in Figure 7.11.

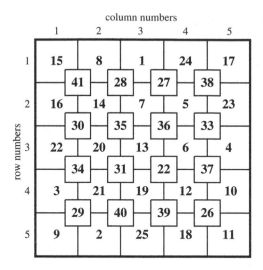

Figure 7.11   Franklin's "four plus five equals nine" magic square.

4	26	50	15	37
48	13	40	2	29
38	5	27	46	16
25	49	14	41	3
17	39	1	28	47

6	33	21	42
44	19	31	8
43	20	32	7
9	30	18	45

11	34	24
36	23	10
22	12	35

(a)   (b)   (c)

Figure 7.12   Sequence of three magic squares.

7.7   Sequence of Three Magic Squares

The numbers $1, 2, \ldots, 50$ have been distributed among the three magic squares shown in Figure 7.12 (Heinz [95]).

The three magic squares in Figure 7.12 are perfect but not standard. The magic constant for the magic square in Figure 7.12(a)

TABLE 7.1 Distribution of the numbers by the squares in Figure 7.12.

5×5	4×4	3×3
1, 2, 3, 4, 5,	6, 7, 8, 9,	10, 11, 12,
13, 14, 15, 16, 17,	18, 19, 20, 21,	22, 23, 24,
25, 26, 27, 28, 29,	30, 31, 32, 33,	34, 35, 36
37, 38, 39, 40, 41,	42, 43, 44, 45	
46, 47, 48, 49, 50		

is 69. For the magic square in Figure 7.12(b), it is 102. For the magic square in Figure 7.12(c), it is 132.

The distribution of the numbers by the squares is shown in Table 7.1.

7.8   Magic Square of Square Numbers

Here, Figure 7.13 is a similar but different magic square of squares that comes from our friend Allan W. Johnson Jr. of Arlington, Virginia [33]. In this case the squared numbers produce a square that is magic when added (rather than when multiplied), to give the magic sum of 93,025. And to add to the fun, the magic sum is also a square, the square of 305.

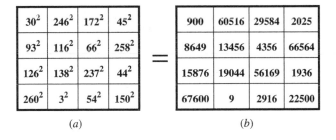

(a)   (b)

Figure 7.13   A magic square of square numbers.

Note that the square taken as the base is not magic. The first row, for example, adds up $30 + 246 + 172 + 45 = 439$, and the second row adds up $93 + 116 + 66 + 258 = 533$.

7.9   Power Magic Square

In mathematics parlance, a magic square is a square array of numbers with some properties. Since a square matrix is also a square array of numbers, a magic square can be represented as a square matrix, which allows the use of the algebra of matrices, such as multiplication and exponentiation (square, cube, etc.), to obtain a new magic square.

By *power magic square* it is understood that the power (square, cube, etc.) of the $n \times n$ matrix representation is obtained from the array of numbers in the magic square. For example, take the Lo Shu 3×3 magic square, Figure 7.14(a) (Swetz [70]). Its matrix representation is shown in Figure 7.14(b). Note that the matrix representation contains the same numbers in the same positions as in the square. Since the square representation is magic, the matrix representation is also magic.

The square of the 3×3 square matrix representation, Figure 7.14(b), of the 3×3 Lo Shu magic squares, Figure 7.14(a),

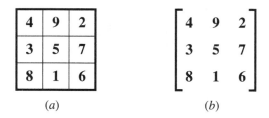

(a)   (b)

Figure 7.14   (a) Lo Shu magic square in traditional form; (b) Lo Shu magic square in matrix form.

is indicated in equation (7.18).

$$
\begin{bmatrix} 4 & 9 & 2 \\ 3 & 5 & 7 \\ 8 & 1 & 6 \end{bmatrix}^2 = \begin{bmatrix} 4 & 9 & 2 \\ 3 & 5 & 7 \\ 8 & 1 & 6 \end{bmatrix} \times \begin{bmatrix} 4 & 9 & 2 \\ 3 & 5 & 7 \\ 8 & 1 & 6 \end{bmatrix}
$$

$$
= \begin{bmatrix} 59 & 83 & 83 \\ 83 & 59 & 83 \\ 83 & 83 & 59 \end{bmatrix} \qquad (7.18)
$$

The product of two matrices, shown in lowercase letters in equation (7.19).

$$
\begin{bmatrix} a & b & c \\ d & e & f \\ g & h & i \end{bmatrix} \times \begin{bmatrix} m & n & o \\ p & q & r \\ s & t & u \end{bmatrix} = \begin{bmatrix} A & B & C \\ D & E & F \\ G & H & I \end{bmatrix} \qquad (7.19)
$$

gives the matrix shown in capital letters according to the following rule.

To obtain the number F located at the intersection of the second row and third column, multiply each number in the second row, that is $[d\,e\,f]$, by the numbers in the third column of the second matrix, which are $[o\,r\,u]$, and then adding the products obtained as follows:

$$
F = \begin{bmatrix} d & e & f \end{bmatrix} \times \begin{bmatrix} o & r & u \end{bmatrix} = \left[(d \times o) + (e \times r) + (f \times u) \right].
$$

$$
(7.20)
$$

Equation (7.18) shows the product of two matrices. For example, to obtain the number $F = 83$, located on the intersection of the second row and third column, take the second row $[d\,ef] = [3\,5\,7]$ of the first term of the product and the third column $[o\,r\,u] = [2\,7\,6]$ of the second term of the product. The resulting product is evaluated as follows:

$$
(3 \times 2) + (5 \times 7) + (7 \times 6) = 6 + 35 + 42 = 83. \qquad (7.21)
$$

The number $A = 59$ in equation (7.18), located at the intersection of the first row and first column, is obtained by multiplying the first row $[a\,b\,c] = [4\,9\,2]$ of the first matrix and the first column $[m\,p\,s] = [4\,3\,8]$ of the second matrix, which gives

$$(4 \times 4) + (9 \times 3) + (2 \times 8) = 16 + 27 + 16 = 59. \qquad (7.22)$$

The computation of the square of the matrix, that is, multiplying by itself, gives another magic square.

Although the rows and columns in the square array obtained in equation (7.18) adds up to the same number, that is 225, which is the square of the magic number of a 3×3 magic square, that is $(15^2 = 225)$, the square array is not a standard magic square because the diagonals are not magic. Swetz [68, p. 2] calls it a "*semi-magic*" square.

The multiplication of the matrix representation of the Lo Shu magic square, which is shown in Figure 7.14(b), by the resultant matrix in equation (7.18), gives the cube of the matrix representation of the Lo Shu magic square, which is exhibited in equation (7.23).

$$\begin{bmatrix} 4 & 9 & 2 \\ 3 & 5 & 7 \\ 8 & 1 & 6 \end{bmatrix}^3 = \begin{bmatrix} 4 & 9 & 2 \\ 3 & 5 & 7 \\ 8 & 1 & 6 \end{bmatrix} \times \begin{bmatrix} 4 & 9 & 2 \\ 3 & 5 & 7 \\ 8 & 1 & 6 \end{bmatrix}$$

$$\times \begin{bmatrix} 4 & 9 & 2 \\ 3 & 5 & 7 \\ 8 & 1 & 6 \end{bmatrix}$$

$$= \begin{bmatrix} 4 & 9 & 2 \\ 3 & 5 & 7 \\ 8 & 1 & 6 \end{bmatrix} \times \begin{bmatrix} 59 & 83 & 83 \\ 83 & 59 & 83 \\ 83 & 83 & 59 \end{bmatrix}$$

$$= \begin{bmatrix} 1149 & 1029 & 1197 \\ 1137 & 1125 & 1077 \\ 1053 & 1221 & 1101 \end{bmatrix} \qquad (7.23)$$

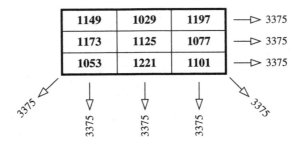

Figure 7.15   The cube of the Lo Shu magic square.

Note that the cube of the matrix representation of the Lo Shu magic square is a matrix with the properties of a perfect magic square since the sum of each number in each row, column, and diagonal adds up to the magic sum 3,375, as Figure 7.15 shows.

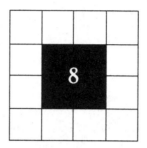

Exotic Magic Squares

8.1 Antimagic Squares

Heterosquares

Because a magic square requires that the rows, columns, and diagonals yield the magic constant, there would inevitably spring up some contrary-minded joker who would devise a square in which all these lines yield different sums. One such mind twister was puzzle master Sam Lloyd, who dreamed up a square like this in the nineteenth century. These squares are given the

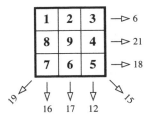

Figure 8.1 Order 3 heterosquare.

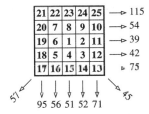

Figure 8.2 Order 5 heterosquare.

name *heterosquares*, and one is shown in Figure 8.1 (Heinz and Hendricks [22]).

The heterosquare of order 5 in Figure 8.2 was constructed according to the following rule:

1. place the number 1 in the central cell,
2. from the central cell move to the next cell on the right and place the number 2,
3. from the cell with the number 2, rotate clockwise and place the numbers from 3 through 9, consecutively, on each cell.
4. move to the cell on the right of the cell with the number 9, and place the number 10.

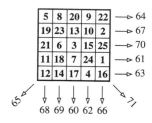

Figure 8.3   Order 5 antimagic square.

5. rotate clockwise, placing the numbers from 11 through 25 on each cell, consecutively. The number 25 will be located on the last cell of the first row.

This procedure is suggested by Heinz and Hendricks [23].

Antimagic Squares

In a special case, when not only are different sums obtained but the sums also form a consecutive series, these are termed *antimagic squares* (Heinz and Hendricks [23]). Figure 8.3 is an order 5 antimagic square where the sums make up a sequence of consecutive integers from 60 to 71.

A math graduate student at the University of Winnipeg, Canada, did a summer research project on antimagic squares and reported that whereas there are 880 magic squares of order 4, there are 299,710 antimagic squares. Regarding order 5 squares, he reported 34,473,153 magic squares but an unknown number of antimagic squares. He developed several methods of constructing these squares for both even and odd orders. For more information about antimagic squares, see Lindon [43], Madachy [47], and John Cormie and Václav Linek's excellent Antimagic Squares Web site [78].

Heterosquares with Prime Numbers

Heinz [94] shows two magic squares constructed with prime numbers by Rivera [106]. The heterosquare in Figure 8.4(a) has nine

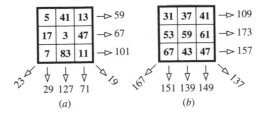

Figure 8.4 Prime numbers heterosquare.

prime numbers such that the sum of numbers along each row, column, and diagonal equals different prime numbers. In addition, the sum of all nine numbers in the square equals the prime number 227.

Figure 8.4(b) shows a heterosquare with the same properties as the one in Figure 8.4(a). It is made of consecutive prime numbers, that is, with the numbers 31, 37, 41, 43, 47, 53, 59, 61, and 67. The sum of each number in all nine cells equals 439, which is also a prime number.

8.2   Reversible Magic Squares

Reversible magic squares are pairs of magic squares that are mirror images of each other, including the digits of numbers in the cells. For example, the cell in the first row, first column, in the square in Figure 8.5(a) contains the number 15. The mirror image of this cell is the cell located in the first row and fourth column of the square in figure 8.5(b), which contains the number 51. The reverse order of the digits in the number 15 is 51, which is the number located in the cell in the first row, fourth column, of the second square.

The magic constant is the same for both squares. For example, the sum of each number in the first row of the first square gives $15 + 94 + 36 + 97 = 242$. The sum of each number in the first row of the second square is $79 + 63 + 49 + 51 = 242$. See Heinz and Hendricks [24].

15	94	36	97
96	37	91	18
93	16	98	35
38	95	17	92

(a)

79	63	49	51
81	19	73	69
53	89	61	39
29	71	59	83

(b)

Figure 8.5   Order 4 reversible magic square.

8.3   Multiplication Reversible Magic Squares

To add a little spice to the reversible magic square broth, R. B. Edwards, an amateur magician from Rochester, New York, concocted a 6 × 6 perfect addition magic square, shown in Figure 8.6, with an additive magic number equal to 1,355. See Pickover [56] and Madachy [46].

The middle 4 × 4 subsquare in the 6 × 6, Figure 8.6, is a multiplication magic square, that is, the product of each number in each row, column, or diagonal gives the same number, 401,393,664, as Figure 8.7 illustrates. For example, the first row gives

$$408 \times 336 \times 244 \times 12 = 401{,}393{,}664, \qquad (8.1)$$

and the second column gives

$$336 \times 24 \times 488 \times 102 = 401{,}393{,}664. \qquad (8.2)$$

The 4 × 4 multiplication magic square in Figure 8.7 is also a multiplicative reverse magic square, that is, the square constructed by reversing each number is also a multiplicative magic square, as Figure 8.8 shows. The multiplicative magic constant of the reversed square is 4,723,906,824.

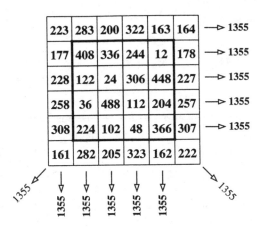

Figure 8.6 Magic square of order 6 with a multiplication reversible subsquare of order 4.

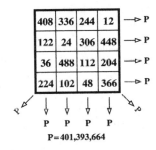

Figure 8.7 Multiplication reversible magic square of order 4.

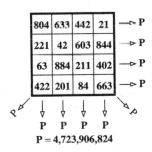

Figure 8.8 Reverse of the multiplication reversible magic squares of order 4.

8.4   Forward-Backward Magic Squares

A surprising relationship was revealed by R. Holmes of London ([30]) for the sum of the squares of the numbers in the rows, columns, and diagonals of the magic square of 3 in Figure 8.9.

The numbers in each row in the magic square in Figure 8.9 can be considered as written, in base ten, using the positional representation, where the number located in the third column of the row has the least significant value. For example, the number represented by the first row is 618. Then obtain the sum of the squares of the numbers obtained from the rows, the columns, and the diagonals of the magic square of order 3 in Figure 8.9, that is,

$$618^2 + 753^2 + 294^2 = 381,924 + 567,009 + 86,436$$
$$= 1,035,369. \tag{8.3}$$

When you reverse the direction, that is, consider the number located in the first column as the least significant, the sum of the squares of the numbers writes

$$816^2 + 357^2 + 492^2 = 655,856 + 127,449 + 242,064$$
$$= 1,035,369. \tag{8.4}$$

6	1	8
7	5	3
2	9	4

Figure 8.9   Magic square of order 3.

That's strange: they produce the same number. Let's try the columns and repeat the process.

$$672^2 + 159^2 + 834^2 = 451{,}584 + 25{,}281 + 695{,}556$$
$$= 1{,}172{,}421. \tag{8.5}$$

In reverse,

$$276^2 + 951^2 + 438^2 = 76{,}176 + 904{,}401 + 191{,}844$$
$$= 1{,}172{,}421. \tag{8.6}$$

Even stranger, and we still have to test the diagonals, the main diagonals, and the pan diagonals. The pan diagonals are the so-called wrap-around diagonals. Both types of diagonals appear as straight lines when the magic square is repeated on the side or on the top of the first square, as in the Figure 8.10. The numbers being considered are indicated by the gray stripes.

$$654^2 + 132^2 + 879^2 = 427{,}716 + 17{,}424 + 772{,}641$$
$$= 1{,}217{,}781. \tag{8.7}$$

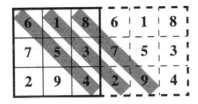

Figure 8.10   Square repeated on the right and lines parallel to the main diagonal.

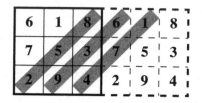

Figure 8.11   Square repeated on the right and lines parallel to the secondary diagonal.

The reverse:

$$456^2 + 231^2 + 978^2 = 207,936 + 53,361 + 956,484$$
$$= 1,217,781. \tag{8.8}$$

Peculiar indeed. See how it works for the secondary diagonal in Figure 8.11.

$$852^2 + 639^2 + 174^2 = 725,904 + 408,321 + 30,276$$
$$= 1,164,501. \tag{8.9}$$

In reverse:

$$258^2 + 936^2 + 471^2 = 66,564 + 876,096 + 221,841$$
$$= 1,164,501. \tag{8.10}$$

By golly, it did it again.

Let's see how it does with the square repeated above, Figure 8.12(a).

$$654^2 + 213^2 + 798^2 = 427,716 + 45,369 + 636,804$$
$$= 1,109,889. \tag{8.11}$$

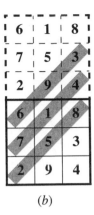

(a)   (b)

Figure 8.12 Square repeated above. (a) Lines parallel to the main diagonal; (b) Lines parallel to the secondary diagonal.

For the reverse,

$$456^2 + 312^2 + 897^2 = 207{,}936 + 97{,}344 + 804{,}609$$
$$= 1{,}109{,}889. \qquad (8.12)$$

Darned if it didn't do it again. Now one more try, Figure 8.12(b):

$$396^2 + 417^2 + 852^2 = 156{,}816 + 173{,}889 + 725{,}904$$
$$= 1{,}056{,}609. \qquad (8.13)$$

The reverse:

$$693^2 + 714^2 + 258^2 = 480{,}249 + 509{,}796 + 66{,}564$$
$$= 1{,}056{,}609. \qquad (8.14)$$

That does it!

We must conclude, therefore, that this magic square works both forward and backward as well.

8.5   Concentric (Bordered) Magic Squares

Figure 8.13 is an order 12 perfect magic square, and more. It has four perfect magic squares nested inside of it. If you peel off the outside border you have a 10 × 10 magic square, and, in turn, peeling off the following borders you have an 8 × 8, a 6 × 6, and a 4 × 4 magic square. The magic constant for the 12 × 12 is 870, for the 10 × 10 is 725, for the 8 × 8 is 580, for the 6 × 6 is 435, and for the 4 × 4 is 290.

It is of interest to note that the sum of the numbers on each diagonal of each magic square equals the sum of the numbers along the same diagonal of the previous magic square plus 145. Analogously, the sum of the numbers on each row/column of each magic square equals the sum of the numbers along the same row/column of the previous magic square plus 145 (Heinz [96]).

1	143	142	4	5	139	138	8	9	135	134	12
13	23	121	120	119	27	29	31	113	112	30	132
131	117	41	103	102	44	45	99	98	48	28	14
130	105	96	55	89	88	59	84	60	49	40	15
129	39	95	87	65	79	78	68	58	50	106	16
128	107	51	62	76	70	71	73	83	94	38	17
18	37	93	82	72	74	75	69	63	52	108	127
19	36	53	64	77	67	66	80	81	92	109	126
125	35	54	85	56	57	86	61	90	91	110	20
21	111	97	42	43	101	100	46	47	104	34	124
22	115	24	25	26	118	116	114	32	33	122	123
133	2	3	141	140	6	7	137	136	10	11	144

Figure 8.13   A 12 × 12 bordered magic square.

8.6   Universal Magic Squares

This crazy square, shown in Figure 8.14 and made up entirely of eights and ones, gives the magic sum of 19,998 for rows, columns, and diagonals in all directions, upside down or when rotated in all directions, and reflected in a mirror. It is pandiagonal, so six complementary diagonal pairs and sixteen 2 × 2 squares give the magic sum. See Heinz and Hendricks [25].

8.7   Annihilation Magic Squares

Annihilation magic squares are squares in which the rows, columns, and diagonals add up to zero, as in Figures 8.15 and Figure 8.16. These bits of nothing at all come to us from the good graces of puzzle man Ivan Moscovich [51].

8818	1111	8188	1881
8181	1888	8811	1118
1811	8118	1181	8888
1188	8881	1818	8111

Original

1118	8181	1888	8811
8888	1811	8118	1181
8111	1188	8881	1818
1881	8818	1111	8188

Rotated 180 degrees

1881	8818	1111	8188
8111	1188	8881	1818
8888	1811	8118	1181
1118	8181	1888	8811

Reflected horizontal

1188	8881	1818	8111
1811	8118	1181	8888
8181	1888	8811	1118
8818	1111	8188	1881

Reflected vertical

Figure 8.14   An order 4 universal magic square.

8	−7	−6	5
−4	3	2	−1
1	−2	−3	4
−5	6	7	−8

Figure 8.15   An order 4 annihilation magic square.

32	-31	-30	29	28	-27	-26	25
-24	23	22	-21	-20	19	18	-17
16	-15	-14	13	12	-11	-10	9
-8	7	6	-5	-4	3	2	-1
1	-2	-3	4	5	6	-7	8
-9	10	11	-12	-13	14	15	-16
17	-18	-19	20	21	-22	-23	24
-25	26	27	-28	-29	30	31	-32

Figure 8.16   An order 8 annihilation magic square.

8.8   Palindromic Magic Squares

A palindrome is a word or number that reads the same forward or backward. A sentence made up of palindromes that reads the same backward and forward is "Able was I ere I saw Elba."

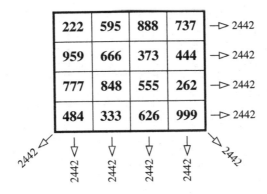

Figure 8.17   An order 4 palindromic magic square.

A palindrome is also an integer that reads the same way forward and backward, for example, 131, 11011, and so on.

Magic square enthusiast Allan W. Johnson Jr. produced the palindromic magic square shown in Figure 8.17. It has a magic sum of 2,442, which is also a palindrome (Johnson [32]).

8.9   Concatenation Magic Squares

Magic Product of Concatenation

Still another surprising combination of numbers gives us a group of concatenation magic squares, as we learn from math professor Emmanuel Emanouilidis of Kean University ([15]). But first let us become familiar with concatenation.

Columns Concatenation

Concatenation refers to the linking together of separate parts. Consider Figure 8.18. The concatenation of column C_1 and

$$M = \begin{array}{c} \\ r_1 \\ r_2 \\ r_3 \end{array} \begin{array}{|c|c|c|} \multicolumn{1}{c}{c_1} & \multicolumn{1}{c}{c_2} & \multicolumn{1}{c}{c_3} \\ \hline 18 & 1 & 12 \\ \hline 4 & 6 & 9 \\ \hline 3 & 36 & 2 \\ \hline \end{array}$$

Figure 8.18 Concatenation of an order 3 magic square.

C_2 gives

$$\begin{bmatrix} 18 \\ 4 \\ 3 \end{bmatrix} \text{ concat } \begin{bmatrix} 1 \\ 6 \\ 36 \end{bmatrix} = \begin{bmatrix} 181 \\ 46 \\ 336 \end{bmatrix}, \qquad (8.15)$$

and then computing the product of the concatenations yields $181 \times 46 \times 336 = 2,797,536$.

Performing the concatenation in reverse order, that is, concatenating C_2 and C_1, gives

$$\begin{bmatrix} 1 \\ 6 \\ 36 \end{bmatrix} \text{ concat } \begin{bmatrix} 18 \\ 4 \\ 3 \end{bmatrix} = \begin{bmatrix} 118 \\ 64 \\ 363 \end{bmatrix}, \qquad (8.16)$$

and it gives the same product, $118 \times 64 \times 363 = 2,797,536$.

The concatenation of columns C_2 and C_3 gives the product $112 \times 69 \times 362 = 2,797,536$; the concatenation of columns C_3 and C_2 yields $121 \times 96 \times 236 = 2,797,536$.

But the product of the concatenation of the columns C_1 and C_3 gives $1812 \times 49 \times 32 = 2,841,216$, and the product of the concatenation of the columns C_3 and C_1 is $1218 \times 94 \times 23 = 2,633,316$, which differ from the previous concatenations.

Concatenation of the Rows

The concatenation of rows gives a similar result but with a product value different from the one obtained by concatenating the columns. For example, concatenating rows r_2 and r_3, it follows

$$\begin{bmatrix} 4 & 6 & 9 \end{bmatrix} \text{concat} \begin{bmatrix} 3 & 36 & 2 \end{bmatrix} = \begin{bmatrix} 43 & 636 & 92 \end{bmatrix}, \quad (8.17)$$

and the product is $43 \times 636 \times 92 = 2,516,016$.

When concatenating rows r_1 and r_2 the number 6 must be written as 06, that is

$$\begin{bmatrix} 18 & 1 & 12 \end{bmatrix} \text{concat} \begin{bmatrix} 4 & 06 & 9 \end{bmatrix} = \begin{bmatrix} 184 & 106 & 129 \end{bmatrix}$$
$$(8.18)$$

and the product is $184 \times 106 \times 129 = 2,516,016$. Similarly for the concatenation of rows r_3 and r_2, the numbers 4 and 9 must be written as 04 and 09, therefore $304 \times 366 \times 209 = 23,254,176$, which is different from the product of the concatenation of rows r_2 and r_3.

Magic Sum of Concatenation

Concatenation of the Columns

The concatenation of the columns of the square in Figure 8.19 is performed analogously to equations (8.15) and (8.16) for the square shown in Figure 8.18. But instead of the product of

	c_1	c_2	c_3
r_1	2	9	4
r_2	7	5	3
r_3	6	1	8

$$M =$$

Figure 8.19   Concatenation of an order 3 magic square.

the resultant concatenation, it is evaluated as the sum of each concatenation.

The concatenation of any two columns, including with itself, gives a total of six possible cases. Adding the respective numbers in each concatenation gives 165 for all cases.

C_1C_1	C_1C_2	C_1C_3	C_2C_1	C_2C_2	C_2C_3	C_3C_1	C_3C_2	C_3C_3
22	29	24	92	99	94	42	49	44
77	75	73	57	55	53	37	35	33
66	61	68	16	11	18	86	81	88
165	165	165	165	165	165	165	165	165

Concatenation of the Rows

The same sum is obtained adding the numbers after the concatenation of any two rows, as shown below.

$$R_1 R_1 \longrightarrow 22 + 99 + 44 = 165$$
$$R_1 R_2 \longrightarrow 27 + 95 + 43 = 165$$
$$R_1 R_3 \longrightarrow 26 + 91 + 48 = 165$$

$$R_2 R_1 \longrightarrow 72 + 59 + 34 = 165$$
$$R_2 R_2 \longrightarrow 77 + 55 + 33 = 165$$
$$R_2 R_3 \longrightarrow 76 + 51 + 38 = 165$$

$$R_3 R_1 \longrightarrow 62 + 19 + 84 = 165$$
$$R_3 R_2 \longrightarrow 67 + 15 + 83 = 165$$
$$R_3 R_3 \longrightarrow 66 + 11 + 88 = 165$$

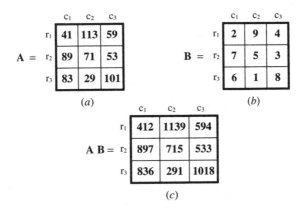

Figure 8.20   (a) Magic square with prime numbers; (b) Magic square in Figure 8.19; (c) Concatenation of two magic squares.

Magic Square Concatenation

If you concatenate any order 3 magic square with the magic square shown in Figure 8.19, the resulting square will be magic.

Take the magic square made up of nine prime numbers and a magic sum of 213, Figure 8.20(a), and the magic square shown in Figure 8.19. When concatenated, as shown in Figure 8.20(a) and (b), it will give the square in Figure 8.20(c) with a magic sum of 2,145.

The cell r_2, c_1 in Figure 8.20(c), for example, was obtained from the concatenation of the value in the cell r_2, c_1 in Figure 8.20(a) (which is 89), with the number in the cell r_2, c_1 in Figure 8.20(b). After the concatenation we obtain 89 7 = 897 for the cell r_2, c_1 in Figure 8.20(c).

8.10   Alphamagic Squares

Perhaps the most unusual and unexpected new addition to the family of magic squares is the invention of the ingenious Lee

5	22	18
28	15	2
12	8	25

Figure 8.21   Initial alphamagic square.

five	twenty-two	eighteen
twenty-eight	fifteen	two
twelve	eight	twenty-five

Figure 8.22   Numeric alphamagic written in words.

Sallows, a British electronics engineer working at the University of Nijmegen in the Netherlands and an expert in word games.

Rather than try to explain alphamagic squares, it is simpler to show one, as in Figure 8.21, an order 3 square with a magic sum of 45 (Stewart [66]).

For each cell, spell the number in that cell and place that word in that cell, as in Figure 8.22.

Now, for each cell, count the number of letters in each word and place that number in a new square, as in Figure 8.23, and you have an entirely new magic square with a magic sum of 21.

Sallows has demonstrated alphamagic for languages other than English, including French, German, Welsh, and even Swahili.

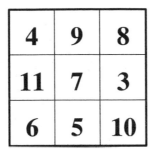

Figure 8.23   A new 3 × 3 magic square.

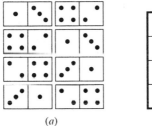

Figure 8.24   (*a*) A 4 × 4 Latin square with dominos; (*b*) The same Latin square with numbers.

8.11   Domino Latin Squares

Figure 8.24(a) shows a 4 × 4 Latin square using dominos. Figure 8.24(b) shows the same Latin square with numbers (Heinz and Hendricks [21]).

8.12   Magic Hexagonal Tiles

The numbers 1, ..., 19 in Figure 8.25 are distributed on hexagonal cells, which forms a hexagonal honeycomb shape (Madachy [48]).

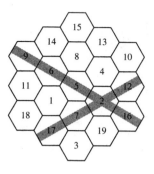

Figure 8.25   Magic hexagon.

The sum of each number in any straight line of hexagonal cells edge-joined equals the magic constant 38. For example, as shown in the shaded lines in Figure 8.25, the addition of each number gives

$$9 + 6 + 5 + 2 + 16 = 38, \qquad (8.19)$$

and

$$17 + 7 + 2 + 12 = 38. \qquad (8.20)$$

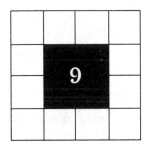

Other Magic Squares

9.1 Associated Magic Squares

In a sequence $\{1, 2, \ldots, n^2\}$ of consecutive numbers, where n is the order of a magic square, any pair of numbers is said to be complementary if their sum equals the sum of the first and last number of the sequence, that is, $n^2 + 1$. For example, in the case of magic squares of order 5, the pairs $(12, 14)$ and $(6, 20)$ are complementary with respect to the sequence $\{1, 2, \ldots, 5^2\}$, where the first is 1 and the last is 25. This implies $n^2 + 1 = 5^2 + 1 = 26$; therefore, $12 + 14 = 6 + 20 = 5^2 + 1$.

16	3	2	13
5	10	11	8
9	6	7	12
4	15	14	1

(a)

16			13
	10	11	
	6	7	
4			1

(b)

Figure 9.1   (a) Dürer-Franklin 4 × 4 magic square; (b) Associated pairs with respect to the center of the square.

A magic square is associated if the numbers on each two cells symmetric with respect to the center are complementary. If the square is even, it must be symmetric with respect to the center of the square.

For example, the Dürer-Franklin 4 × 4 magic square is associated, as Figure 9.1 shows. The sum of each number on any pair equally distant from the center, equals $n^2 + 1 = 4^2 + 1 = 17$, for example (16, 1) are complementary numbers, as well as (6, 11), (13, 4), and (7, 10).

If the order of the square is odd, then the pairs are symmetrical with respect to the central cell; for example, as for the 5 × 5 magic square shown in Figure 9.2(a) (Descombes [11]). The sum of each pair must equal $5^2 + 1 = 26$. For example, along the main diagonal (23, 3), (18, 8); along the secondary diagonal (11, 15), (12, 14); along the middle row (17, 9), (5, 21); and along the middle column (19, 7), (1, 25) are complementary, and these pairs are shown separately in 9.2(b).

For an odd-order magic square, the sum of the first and last terms equals the double of the middle term of the sequence, which is the value that must be stored on the middle cell. For example, for the magic square in Figure 9.2, the middle term is given by $(5^2 + 1)/2 = 13$, which is the value assigned to the central cell.

23	6	19	2	15
10	18	1	14	22
17	5	13	21	9
4	12	25	8	16
11	24	7	20	3

(a)

23		19		15
	18	1	14	
17	5	13	21	9
	12	25	8	
11		7		3

(b)

Figure 9.2   (a) A 5 × 5 magic square; (b) Associated pairs with respect to the center cell of the square.

71	89	17
5	59	113
101	29	47

Figure 9.3   Symmetrical magic square with prime numbers.

9.2   Symmetrical Magic Squares

The definition of complementary pairs also applies for a sequence of nonconsecutive numbers, as in the sequence {5, 17, 29, 47, 59, 71, 89, 101, 113}.

If the sequence of numbers filling a magic square is made of nonconsecutive numbers, then the magic square is called *symmetrical*, which is the case of the square shown in Figure 9.3. It is made up of the sequence of prime numbers {5, 17, 29, 47, 59, 71, 89, 101, 113}. The middle number, which occupies the center cell, is given by $(5 + 113)/2 = 59$. The sum of the numbers on each pair symmetric with respect to the cell with 59 gives twice 59, that is, 118.

16	3	2	13
5	10	11	8
9	6	7	12
4	15	14	1

(a)

1	14	15	4
12	7	6	9
8	11	10	5
13	2	3	16

(b)

Figure 9.4   (a) Dürer-Franklin's 4 × 4 magic square;
(b) Dürer-Franklin's complementary magic square.

9.3   Complementary Magic Squares

Two magic squares are complementary if the cell of one square has the value x, then the cell at the same position on the other square has the value $(n^2 + 1) - x$, where n is the order of the magic square.

Figure 9.4(a) shows the Dürer-Franklin 4 × 4 magic square and Figure 9.4(b) shows the complementary magic square derived from Dürer-Franklin's square of order 4. In this case $n = 4$, therefore, $n^2 + 1 = 4^2 + 1 = 17$. Each number on each cell in the square in Figure 9.4(b) was obtained by subtracting each number x in the cell at similar position in the square in Figure 9.4(a) from 17. For example, the cell at the position (row 2, column 1) in the square in Figure 9.4(a) contains the number 5. Therefore, the cell at the position (row 2, column 1) in the square in Figure 9.4(b) must contain the number $17 - 5 = 12$.

If the complementary square coincides with the original, after one or more rotations, the complementary square is called *self-complementary*. This is the case of Dürer-Franklin, which can be seen from Figure 9.5.

9.4   Magic Square with Consecutive Pairs of Numbers

Mannke [49] created a perfect 8 × 8 magic square (Figure 9.6) with a magic constant of 260 and with the additional property that consecutive pairs of numbers are in adjacent squares. In addition, the first 32 numbers can be joined by one line.

13	8	12	1
2	11	7	14
3	10	6	15
16	5	9	4

(a)

16	3	2	13
5	10	11	8
9	6	7	12
4	15	14	1

(b)

Figure 9.5   (a) Dürer-Franklin's 4 × 4 magic square after the first clockwise rotation of 90°; (b) Dürer-Franklin's complementary magic square after one more clockwise rotation of 90°.

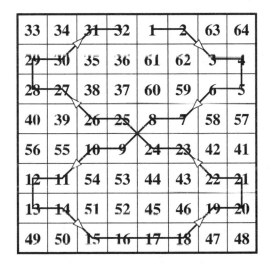

Figure 9.6   Magic square with consecutive pairs of numbers.

9.5   Latin Square Made Up of Magic Squares

Gridgeman [17] presents a Latin square of order 16 partitioned into 16 magic squares of order 4.

A Latin square is an $n \times n$ grid filled with numbers so that the numbers 1 through n occur exactly once in each

1	8	13	12	2	11	14	7	4	5	16	9	3	10	15	6
15	10	3	6	16	5	4	9	14	11	2	7	13	8	1	12
4	5	16	9	3	10	15	6	1	8	13	12	2	11	14	7
14	11	2	7	13	8	1	12	15	10	3	6	16	5	4	9
12	13	8	1	6	15	10	3	9	16	5	4	7	14	11	2
7	2	11	14	9	4	5	16	6	3	10	15	12	1	8	13
9	16	5	4	7	14	11	2	12	13	8	1	6	15	10	3
6	3	10	15	12	1	8	13	7	2	11	14	9	4	5	16
13	12	1	8	14	7	2	11	16	9	4	5	15	6	3	10
3	6	15	10	4	9	16	5	2	7	14	11	1	12	13	8
16	9	4	5	15	6	3	10	13	12	1	8	14	7	2	11
2	7	14	11	1	12	13	8	3	6	15	10	4	9	16	5
8	1	12	13	10	3	6	15	5	4	9	16	11	2	7	14
11	14	7	2	5	16	9	4	10	15	6	3	8	13	12	1
5	4	9	16	11	2	7	14	8	1	12	13	10	3	6	15
10	15	6	3	8	13	12	1	11	14	7	2	5	16	9	4

Figure 9.7   Latin square of magic squares.

row and each column of the square. In the square in Figure 9.7 the order 16 Latin square is made up of 16 4 × 4 squares that are perfect magic squares.

9.6   Franklin Lightning Baseball: Three Strokes and You're Out

Do magic squares have to be square? That sounds like a simple question. They are not required to be square; they are square in name only. They can be magic rectangles, magic triangles, and other shapes, according to the imagination and skill of their inventor. Let's start off with an unusual one that honors our hero, Benjamin Franklin. Its inventor, Edward W. Shineman Jr. of New York City [64], calls it *magic lightning* (Figure 9.8).

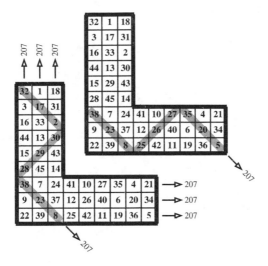

Figure 9.8   Magic lightning.

Three vertical strokes:

$$(32 + 17 + 2) + (30 + 29 + 28) + (38 + 23 + 8) = 207. \quad \textbf{(9.1)}$$

Three horizontal strokes = 207:

$$(38 + 23 + 8) + (25 + 26 + 27) + (35 + 20 + 5) = 207. \quad \textbf{(9.2)}$$

Three vertical columns:

$$32 + 3 + 16 + 44 + 15 + 28 + 38 + 9 + 22 = 207,$$
$$1 + 17 + 33 + 13 + 29 + 45 + 7 + 23 + 39 = 207,$$
$$18 + 31 + 2 + 30 + 43 + 14 + 24 + 37 + 8 = 207. \quad \textbf{(9.3)}$$

Three horizontal rows:

$$38 + 7 + 24 + 41 + 10 + 27 + 35 + 4 + 21 = 207,$$
$$9 + 23 + 37 + 12 + 26 + 40 + 6 + 20 + 34 = 207,$$
$$22 + 39 + 8 + 25 + 42 + 11 + 19 + 36 + 5 = 207. \quad (9.4)$$

9.7   Magic Squares with 666 Magic Constant

Devil's Magic Square

The magic square in Figure 9.9 is called the *devil's magic square* because the magic number is 666, which is called the number of the beast in the Book of Revelations in the New Testament. The numbers in this 4×4 square are not in the standard sequence $1, 2, \ldots, 4^2$, but they are such that the sum of each number in each row, column, diagonal, and pandiagonal gives 666.

Figure 9.9 shows the row, column, and diagonal properties (Ondrejka [52]) and Figures 9.10–9.12 show with shading the other ways to obtain the devil's number in this square.

Figure 9.10(a) shows a 2×2 subsquare, adding up 666, that is,

$$169 + 138 + 151 + 208 = 666. \quad (9.5)$$

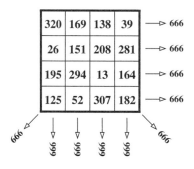

Figure 9.9   A perfect nonstandard magic square.

320	169	138	39
26	151	208	281
195	294	13	164
125	52	307	182

(a)

320	(169)	138	(39)
26	151	208	281
195	(294)	13	(164)
125	52	307	182

(b)

Figure 9.10   Subsquare properties.

320	169	138	39
(26)	151	208	(281)
(195)	294	13	(164)
125	52	307	182

(a)

320	(169)	(138)	39
26	151	208	281
195	294	13	164
125	(52)	(307)	182

(b)

Figure 9.11   Subrectangle properties.

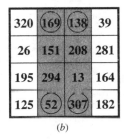

320	169	138	39
26	151	208	281
195	294	13	164
125	52	307	182

Figure 9.12   Four corner properties of the original square.

Figure 9.10(b) shows an example of the four corner property of a 3×3 subsquare, adding up 666, that is,

$$169 + 39 + 164 + 294 = 666. \qquad (9.6)$$

Figure 9.11(a) shows a 2 × 4 subrectangle, whose corners add up 666, that is,

$$26 + 281 + 164 + 195 = 666. \tag{9.7}$$

Figure 9.11(b) shows a 4 × 2 subrectangle, whose corners add up 666, that is,

$$169 + 138 + 307 + 52 = 666. \tag{9.8}$$

Figure 9.12 shows the four corners property of the original square, that is,

$$320 + 39 + 125 + 182 = 666. \tag{9.9}$$

Beastly Magic Square

Another interesting magic square is shown by Heinz [94], who says that he received it from Geest [90] on December 7, 1998.

The magic square in Figure 9.13 was constructed with the first 36 multiples of 6, that is, $6n$ where $n = 2, 3, \ldots, 36$, which expands as $6, 12, 18, \ldots, 216$.

The addition of each number along the third row gives

$$M_6 = 90 + 60 + 102 + 198 + 168 + 48 = 666, \tag{9.10}$$

and along the second column

$$M_6 = 108 + 84 + 60 + 162 + 138 + 114 = 666. \tag{9.11}$$

Franklin's 6 x 6 Magic Square Transformed into a Beastly One

The magic square in Figure 9.15 is beastly because, like the previous square, the magic constant is 666.

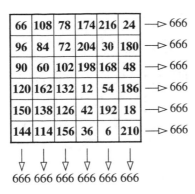

66	108	78	174	216	24	⟶▷ 666
96	84	72	204	30	180	⟶▷ 666
90	60	102	198	168	48	⟶▷ 666
120	162	132	12	54	186	⟶▷ 666
150	138	126	42	192	18	⟶▷ 666
144	114	156	36	6	210	⟶▷ 666

666 666 666 666 666 666

Figure 9.13   A magic square with multiples of 6.

If each cell in any standard 6 × 6 magic square is multiplied by 6, the magic number of the resultant square will be 666. Assume that $a_1 + a_2 + a_3 + a_4 + a_5 + a_6 = M$. The multiplication of each side by p gives

$$p \times (a_1 + a_2 + a_3 + a_4 + a_5 + a_6) = p \times M, \qquad (9.12)$$

which expands as

$$(p \times a_1) + (p \times a_2) + (p \times a_3) + (p \times a_4)$$
$$+ (p \times a_5) + (p \times a_6) = p \times M. \qquad (9.13)$$

Therefore, it is valid for any p. The magic constant of an order 6 magic square, is $M = 111$; when multiplied by $p = 6$ this gives $M = 666$. This permits the transformation of Franklin's 6 × 6 magic square shown in Figure 9.14, into another, whose magic constant equals 666, as shown in Figure 9.15.

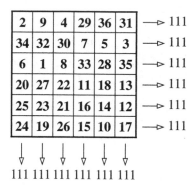

Figure 9.14   Franklin's order 6 magic square.

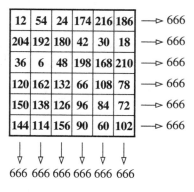

Figure 9.15   A magic square with multiples of 6 obtained from
Franklin's order 6 magic square.

9.8   Pandigital Magic Squares

In a pandigital magic square each number in each cell contains all
ten digits (0, 1, . . . , 8, 9), however the digit zero is not permitted
to occupy the leading (most significant) position.

It should be clear that the words *digit* and *number* have different meaning. A number is composed of digits. For example, the

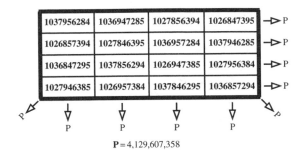

P = 4,129,607,358

Figure 9.16   Pandigital order 4 perfect magic square.

numbers 035, 350, 305, 503, and 530 are composed of the same three digits 0, 3, 5 but represent different quantities. Moreover, the number 035 has a leading zero, and it represents the same quantity as the number 35, which contains two digits. Therefore, if the number must contain ten digits, the zero cannot be in the leading position, because the ten-digit number will represent the same quantity as the number without the leading zero, which is a nine-digit number.

The pandigital magic square shown in Figure 9.16 represents the solution given by Rodolfo Marcelo Kurchan [34] to the following problem: Find the smallest, nontrivial, magic square having n^2 distinct, pandigital integers, and having the smallest, pandigital magic sum.

Note: To save space in Figure 9.16 the 4 × 4 square is presented in rectangular form.

9.9   Magic Square Olympics

From the Internet ([103]), Lies De Sutter, An De Brandt, Katrien De Bruyeker, and Vicky De Marteau inform us of the world records in size of magic squares.

For the largest magic square calculated by hand, in 1990 the prize went to Norbert Behnke of Krefeld, Germany, for a square of order 1,111.

For computer-derived squares, the winners were

1975   Richard Suntag of Pomona, California, with
        105 × 105
1979   Gerolf Lenz of Wuppertal, Germany, with
        501 × 501
1987   Frank Tast and Uli Schmidt of Pforzheim,
        Germany, with 897 × 897; Christian Schaller of
        Munich, Germany, with 1,000 × 1,000; Sven
        Paulus, Ralph Bulling, and Jorg Sutter with
        2,001 × 2,001
1991   Ralf Laue of Leipzig, Germany, with 2,121 × 2,121
1994   Louis Caya of Sainte-Foy, Canada, with
        3,001 × 3,001

Franklin would have been delighted to see how huge magic squares have become since his days, and also to see all the different approaches to the magic number. In this brief review of the magic square curiosities, wherever possible, third- or fourth-order squares are chosen as examples for the sake of simplicity, although in most cases larger order squares are to be found. We have not exhausted all the magic square curiosities: there are inlaid magic squares [29] and infinity magic squares [40] and other novel types in the literature.

9.10   A Magic Square to Show Off

Here's one you can use to show off your magical power to your friends. Figure 9.17 is an order 5 square (Brandreth [7]).

1. Ask one of your friends to select any one of the numbers in the square in Figure 9.17 but not to disclose the number.

2. Ask him to cross off all the numbers that are horizontally, vertically, or diagonally in line with this number.

3. Then request that he does the same with any of the other numbers not crossed off, repeating the process four times until just five numbers are not crossed off.

4. Now ask the friend to add these numbers. Before he tells you the answer, in your superior wisdom, you inform him that the magic sum is 666. You know that because no matter what numbers are in the square he selects, the sum will be 666.

First choice: 248. Then cross the other numbers on the first row and third column. This choice is shown in Figure 9.18(a).

Second choice: 48. Then cross the other numbers on the second row and first column. The addition of this choice to the square is shown in Figure 9.18(b).

Third choice: 127. Then cross the other numbers on the third row and fifth column. The addition of this choice to the square is shown in Figure 9.19(a).

Fourth choice: 130. Then cross the other numbers on the fifth row and second column. The addition of this choice to the square is shown in Figure 9.19(b).

125	191	248	169	116
48	114	171	92	39
136	202	259	180	127
69	135	192	113	60
64	130	187	108	55

Figure 9.17   A square to show off.

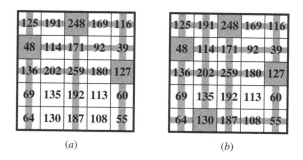

125	191	248	169	116
48	114	171	92	39
136	202	259	180	127
69	135	192	113	60
64	130	187	108	55

(a)

125	191	248	169	116
48	114	171	92	39
136	202	259	180	127
69	135	192	113	60
64	130	187	108	55

(b)

Figure 9.18   Show-off magic square.

125	191	248	169	116
48	114	171	92	39
136	202	259	180	127
69	135	192	113	60
64	130	187	108	55

(a)

125	191	248	169	116
48	114	171	92	39
136	202	259	180	127
69	135	192	113	60
64	130	187	108	55

(b)

Figure 9.19   Show-off magic square.

Figure 9.19(b) shows that there is one number that was not crossed. Therefore, the solution contains the four selected numbers plus the number that was not crossed. The sum of each of these numbers gives

$$248 + 48 + 127 + 130 + 113 = 666. \qquad (9.14)$$

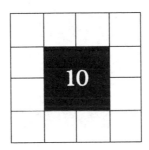

Magic Squares in Art, Design, and Music

10.1 Art and Mathematics

Art Based on Mathematical Symbols

Art is everywhere and mathematics, though less obvious, is also everywhere. Art is in nature, in the human form, in great cathedrals, in Hindu temples, and in Muslim mosques in the abstract patterns used as decoration. In mathematics, art is in lines and angles and curves; in circles, squares, and triangles; in spheres, cubes, cylinders, and ellipsoids; and all those other objects we see.

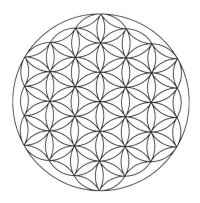

Figure 10.1   Flower of Life.

We have become conscious of the combination of mathematics and art. There is art that is related to math, where mathematical objects like spheres, squares, and cubes are used in drawings and sculptures, and then there is art that is dependent on mathematics, relying on equations and formulations for its visual effects.

The pleasing effect of the former type was recognized by early Egyptians as exemplified by the Flower of Life (Figure 10.1) [88] as seen on the Temple of Osiris 5,000 years ago. Ancient Greeks also recognized the relationship between math and art. Aristotle said, "The mathematical sciences particularly exhibit order, symmetry, and limitation, and these are the greatest forms of the beautiful." Plato and Archimedes are known for the polyhedron solids that appear in the painting of Pacioli by the Renaissance artist Jacapo de Barberi, also of Dürer's *Melencolia*, and Salvador Dalí's *The Sacrament of the Last Supper*. Piet Mondrian obtained striking effects with rectangles of different size and colors.

M. C. Escher, the Dutch artist [86], is well known for his graphic art and his interest in mathematical forms. Among these are tessellations, or tiles, composed of repeating patterns of irregular shapes (Figure 10.2). It is one example of what has been called

Figure 10.2   Escher's horses.

mathematical art and was inspired by mathematician H. S. M. Coxeter. Another example is the ancient Japanese art form of origami, created by folding paper into interesting shapes, which links art and mathematics because the paper folds produce geometric forms and the creases produce bisect angles.

Mathematical art has found its own renaissance today, with a museum in New York devoted entirely to it [89], and many colleges giving courses and symposia on the subject. Books and exhibitions of artists' works can be found everywhere. A search on the Web for the subject of mathematical art yields hundreds of hits. Mathematical art is all around us—in the veins of leaves; in the structure of snowflakes; in patterns of fabrics and wallpaper; and in the design of furniture, autos, and industrial machines.

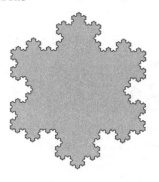

Figure 10.3   Von Koch's snowflake.

Fractals

Most of the art we have known in the past would fall into the category of being related to math. Fractals are examples of a newer type of abstract art that not only gets inspiration from math but also depends on it. The word *fractal* was coined by Benoit Mandelbrot in 1975, but the math began back before the turn of the twentieth century. A fractal has been defined as a rough or fragmented geometric shape that can be subdivided in parts and reduced in size but that remains a copy of the whole. Fractals have contributed to business, industry, and medicine as well as art. Mathematicians contributing to fractal art include Helge von Koch (1904), known for his snowflake (Figure 10.3).

Waclaw Sierpinski is associated with his carpet of squares (Figure 10.4) and also for his triangles. Dr. Mario Markus of the Max Planck Institute for Nutrition, working in computer graphics, devised the Markus-Lyapunov fractals based on the math of the nineteenth-century Russian mathematician Alexander Lyapunov to produce tantalizing visuals (Figure 10.5).

According to Robert L. Devaney of the Department of Mathematics of Boston University, "The Mandelbrot set [Figure 10.6]

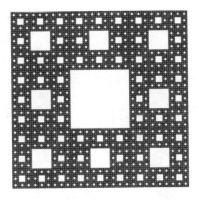

Figure 10.4 Sierpinski carpet.

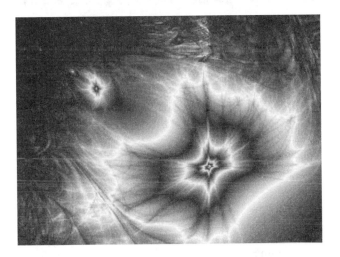

Figure 10.5 Vision of Chaos, Lyapunov fractal [108].

by Benoit Mandelbrot, is one of the most intricate and beautiful images in all mathematics ... most people who have seen this
image have marveled at its geometric intricacy." A comment from
Simon Arthur states, "There are many strange and beautiful sights

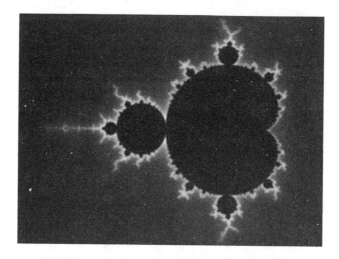

Figure 10.6   The Mandelbrot set.

to see when you explore the Mandelbrot set ranging from the sub-
lime to the psychedelic." This and fractals by others have many
exotic variations.

10.2   Magic Squares and Art

A Web site (www.kandaki.com) under the title *Les Carrés Magiques* (The Magic Squares) gives examples of the work of 45 artists.
Most of their productions use mainly squares, are indeed artisti-
cally creative, and even mathematical. Figure 10.7 shows squares
regularly diminishing in size, yet they give a different, less exact
connotation to the understanding of the term *magic squares*.

Each black square was constructed such that its side equals
a fraction of the side of the cell. The fractions used are
$\{0.8, 0.7, \ldots, 0.1\}$.

Magic squares are created with numbers and derived from
mathematical formulae, so they qualify as mathematics, but do
they qualify as art? Art comes from many sources—from the shape

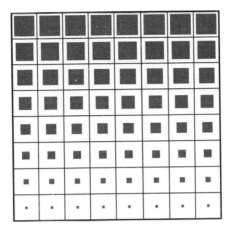

Figure 10.7   The size-reducing squares by Hannes Bürgel.

and color of a flower and from the abstractions of paint drop-
pings on a canvas. Magic square patterns represent a distinct form
of mathematical art. Architect-designer Claude Bragdon (1866–
1946) found magic squares to be a rich source of artistic design for
carpets, curtains, and wall coverings (Figure 10.8).

Bragdon made his philosophical statement: "In mathematics
I seem to have found the source of all ornament whatsoever, and
it was there I decided to plant my metaphysical spade. Ours is the
age of mathematics, it is the magician's wand without which our
workers of magic, be they bankers, engineers, physicists, inventors
could not perform their tricks" [6].

In explaining how he selected magic squares to express math
and art, he said Keats's dictum, "Beauty is truth: truth is beauty,"
gave him the clue. "Because mathematical truth is absolute within
its limits," he declared, "I had only to discover some method of
translation of truth in the mind to beauty in the eye." Magic squares
were selected because "they constitute such a conspicuous instance
of harmony-of-number of mathematical truth."

Figure 10.8   Bragdon's household decorations derived
from magic squares.

To explain his method, Bragdon said, "Every magic square
contains a magic path, discoverable by the numbers in their original
and natural sequence from cell to cell and back again to the original
number. This is called the magic line. Such a line makes of necessity
a pattern, interesting always, and sometimes beautiful as well. Here
is the raw material of ornament: in this way the chasm between
mathematical truth and visible beauty may be bridged."

Bragdon's method, illustrated in Figure 10.9, is what we have
termed the number sequence line. Begin by drawing a line, starting
with the lowest number, proceeding in sequence to the highest,
then return to the first. As Bragdon illustrates in Figures 10.10 and
10.11, one can get different patterns by either taking only odd or
even numbers, or by skipping one, two, or more numbers, then
placing sequences together. In Figure 10.10, the line first proceeds

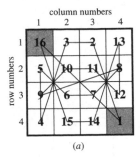
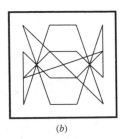

Figure 10.9 (a) Construction of the lines for the Dürer-Franklin 4 × 4 by the Bragdon number sequence method; (b) Pattern obtained from (a).

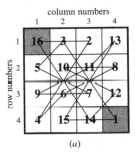
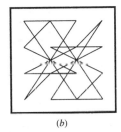

Figure 10.10 (a) Sequence of odd then even numbers {1, 3, 5, ..., 15} and then {2, 4, 6, ..., 16}; (b) Pattern obtained from (a).

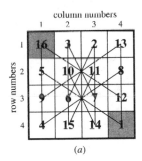
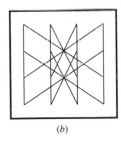

Figure 10.11 (a) Sequences skipping four numbers, odd then even {1, 5, 9, 13}, then {3, 7, 11, 15}, then {2, 6, 10, 14}, then {4, 8, 12, 16}; (b) Pattern obtained from (a).

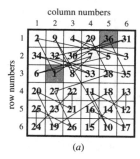
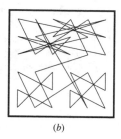

column numbers
1 2 3 4 5 6

(a)   (b)

Figure 10.12   (a) Construction of the lines for the Franklin 6 × 6;
(b) Pattern obtained from (a).

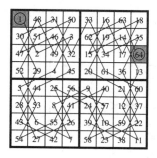

Figure 10.13   Euler's knight's tour on chessboard magic square by
number sequence line method.

with odd numbers 1, 3, 5, etc., then switches to an even number
sequence 2, 4, 6, to 16. In Figure 10.11, the line follows a sequence
skipping two sets of two odd numbers, then two sets of two even
numbers. The patterns obtained with Bragdon's line method are
further illustrated with different squares in Figures 10.12 to 10.16.

Figure 10.12 shows the same line construction and the pattern
for Franklin's order 6 square. A different sequence is followed in
Euler's knight's move on the chessboard with an order 8 magic
square shown with the pattern in Figure 10.13. In these, the starting
and final cells are shaded.

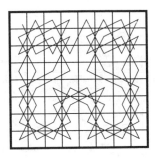

Figure 10.14 Pattern of the Euler's knight's move on a chessboard by number sequence line method.

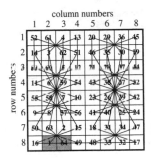

Figure 10.15 Construction of the lines for the first Franklin 8 by number sequence line method, as constructed by the authors.

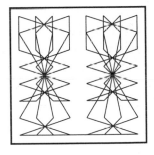

Figure 10.16 Patterns with lines constructed from the first Franklin 8 in Figure 10.15.

column numbers

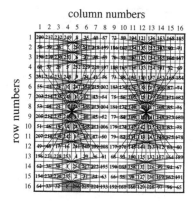

Figure 10.17   Lines for Franklin first 16 by number sequence
method, as constructed by Bragdon.

Franklin made three magic squares of order 8. His first, which
is best known; the second, which he made later; and the third,
which he used as a framework for his magic circle. Figure 10.15
shows his first 8 and the magic line from number 1 to number 64.
In Figure 10.16, we see the pattern of the number sequence line
unencumbered by the distraction of the numbers and shading of
cells 1 and 64. Likewise, Figure 10.16 lets us see the pattern without
the numbers. In earlier discussion, it was mentioned that Franklin
produced two order 16 magic squares that we refer to as the first 16
and the perfect 16 squares. Figure 10.17 shows the magic line for the
first 16, and Figure 10.18 gives the pattern without the numbers.
Similarly, for the perfect 16, Figure 10.19 gives the construction of
the magic line and Figure 10.20 shows just the pattern.

Bragdon did not originate the magic line, but he was the first to
adopt it for the sake of beauty. In so doing, he sometimes modified
it for beauty by curving the lines for delicate effect and by sketching
in lovely designs, as are seen in his work (Figures 10.21 and 10.22).

Magic Constant Line

There is also another way that magic squares may give artistic patterns, called the magic constant line. It is obtained by drawing a

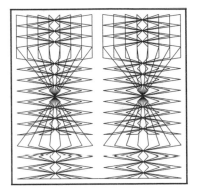

Figure 10.18   Pattern lines for Franklin first 16.

column numbers

Figure 10.19   Construction of the lines for Franklin perfect 16 by the number sequence method, as constructed by the authors.

continuous line through the cells of the magic square, the rows, columns, diagonals, or every combination of numbers that produces the magic constant. Instead of a line, one may shade the cells that give the magic constant, which may be preferred when the numbers are not adjacent. Figure 10.23(a) shows this line method,

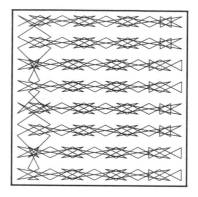

Figure 10.20   Pattern lines for Franklin perfect 16.

and 10.23(b) shows the shading procedure. Figure 10.24 demonstrates the line patterns for all of the ways that the Dürer-Franklin magic square (Dürer square, later made independently by Franklin) will produce the magic constant of 34 for the rows, columns, and diagonals, with the numbered square, top row, center.

There are 40 different ways to obtain the magic constant of 34 with the Dürer-Franklin magic square employing the magic constant lines in Figure 10.24. It shows all these ways for the rows, columns, and diagonals with the numbered square top row, center.

Magic Constant Lines Obtained from the Perfect 16

An example of Franklin's perfect order 16 square with bent-up rows is found in Figure 10.25 and the pattern is in 10.26.

Figure 10.27 shows the rows reversed from Figure 10.25, but this represents not just a reversal but entirely new paths using different numbers to obtain the magic constant. Likewise, Figures 10.29 and 10.31 represent new and different ways to obtain the magic constant. This applies to all figures that follow in this chapter. Figures 10.26, 10.28, 10.30, and 10.32 show the patterns produced without the obstruction of the lines and numbers.

Figure 10.21   Delicate design by curving the lines from
order 3 magic squares.

Figures 10.29 and 10.31 show the bent columns for the
Franklin's perfect order 16 magic square.

Patterns Derived by the Present Authors

Franklin did not exhaust all of the possible patterns that can
produce the magic constant with his wonderful perfect 16 × 16.
Neither have we, but from here to the end of this chapter are

Figure 10.22   Intricate design from curved and straight lines
of order 5 magic square.

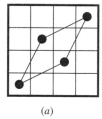

16	3	2	13
5	10	11	8
9	6	7	12
4	15	14	1

(a)   (b)

Figure 10.23   Right-tilted diamond. (a) Magic constant shown
by lines; (b) Magic constant by shading.

original patterns the present authors derived from Franklin's perfect 16 that give the magic constant: Figure 10.33, the magic
diamond; Figure 10.34, the diamond cluster; Figure 10.35, five diamonds; Figure 10.36, a magic Star of David; Figure 10.37–10.40,
opposite double-bent columns; Figures 10.41–10.44, opposite

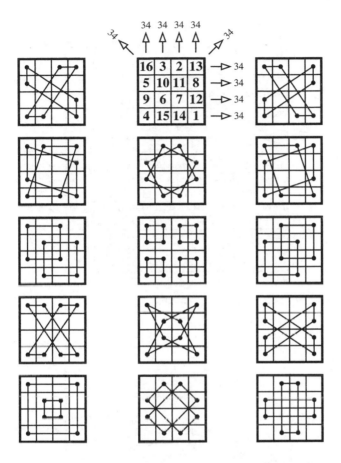

Figure 10.24   Dürer-Franklin 4 × 4 magic square showing magic constant by line method.

quadruple-bent columns; Figures 10.45–10.48, down and up magic candelabras; Figures 10.49–10.50, zig-zag rows; Figures 10.51–10.52, zig-zag columns; Figures 10.53–10.54, interlaced zig-zag rows; Figures 10.55–10.56, interlaced zig-zag columns; and Figures 10.57–10.58, ornamental design with zig-zag rows and column.

column numbers

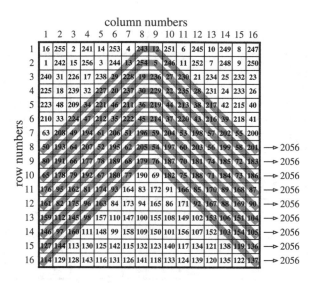

	1	2	3	4	5	6	7	8	9	10	11	12	13	14	15	16	
1	16	255	2	241	14	253	4	243	12	251	6	245	10	249	8	247	
2	1	242	15	256	3	244	13	254	5	246	11	252	7	248	9	250	
3	240	31	226	17	238	29	228	19	236	27	230	21	234	25	232	23	
4	225	18	239	32	227	20	237	30	229	22	235	28	231	24	233	26	
5	223	48	209	34	221	46	211	36	219	44	213	38	217	42	215	40	
6	210	33	224	47	212	35	222	45	214	37	220	43	216	39	218	41	
7	63	208	49	194	61	206	51	196	59	204	53	198	57	202	55	200	
8	50	193	64	207	52	195	62	205	54	197	60	203	56	199	58	201	→ 2056
9	80	191	66	177	78	189	68	179	76	187	70	181	74	185	72	183	→ 2056
10	65	178	79	192	67	180	77	190	69	182	75	188	71	184	73	186	→ 2056
11	176	95	162	81	174	93	164	83	172	91	166	85	170	89	168	87	→ 2056
12	161	82	175	96	163	84	173	94	165	86	171	92	167	88	169	90	→ 2056
13	159	112	145	98	157	110	147	100	155	108	149	102	153	106	151	104	→ 2056
14	146	97	160	111	148	99	158	109	150	101	156	107	152	103	154	105	→ 2056
15	127	144	113	130	125	142	115	132	123	140	117	134	121	138	119	136	→ 2056
16	114	129	128	143	116	131	126	141	118	133	124	139	120	135	122	137	→ 2056

row numbers

Figure 10.25   Franklin bent-up rows.

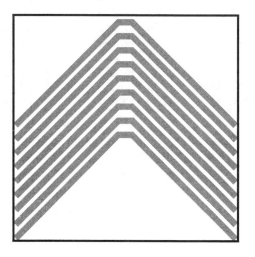

Figure 10.26   Pattern of bent-up rows.

column numbers

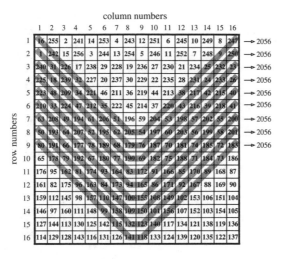

	1	2	3	4	5	6	7	8	9	10	11	12	13	14	15	16	
1	16	255	2	241	14	253	4	243	12	251	6	245	10	249	8	247	→ 2056
2	1	242	15	256	3	244	13	254	5	246	11	252	7	248	9	250	→ 2056
3	240	31	226	17	238	29	228	19	236	27	230	21	234	25	232	23	→ 2056
4	225	18	239	32	227	20	237	30	229	22	235	28	231	24	233	26	→ 2056
5	223	48	209	34	221	46	211	36	219	44	213	38	217	42	215	40	→ 2056
6	210	33	224	47	212	35	222	45	214	37	220	43	216	39	218	41	→ 2056
7	63	208	49	194	61	206	51	196	59	204	53	198	57	202	55	200	→ 2056
8	50	193	64	207	52	195	62	205	54	197	60	203	56	199	58	201	→ 2056
9	80	191	66	177	78	189	68	179	76	187	70	181	74	185	72	183	→ 2056
10	65	178	79	192	67	180	77	190	69	182	75	188	71	184	73	186	
11	176	95	162	81	174	93	164	83	172	91	166	85	170	89	168	87	
12	161	82	175	96	163	84	173	94	165	86	171	92	167	88	169	90	
13	159	112	145	98	157	110	147	100	155	108	149	102	153	106	151	104	
14	146	97	160	111	148	99	158	109	150	101	156	107	152	103	154	105	
15	127	144	113	130	125	142	115	132	123	140	117	134	121	138	119	136	
16	114	129	128	143	116	131	126	141	118	133	124	139	120	135	122	137	

row numbers

Figure 10.27   Franklin bent down rows.

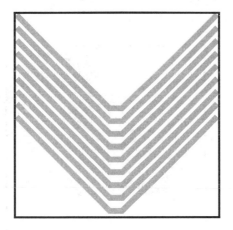

Figure 10.28   Pattern of bent-down rows.

143

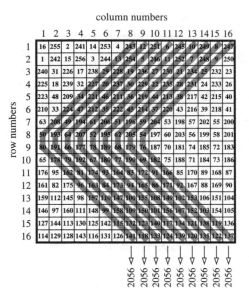

Figure 10.29 Franklin bent-left column.

Figure 10.30 Pattern of bent-left column.

column numbers

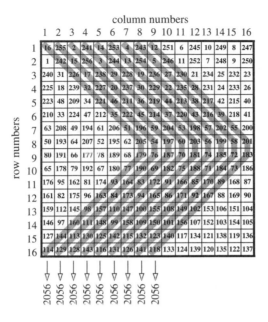

	1	2	3	4	5	6	7	8	9	10	11	12	13	14	15	16
1	16	255	2	241	14	253	4	243	12	251	6	245	10	249	8	247
2	1	242	15	256	3	244	13	254	5	246	11	252	7	248	9	250
3	240	31	226	17	238	29	228	19	236	27	230	21	234	25	232	23
4	225	18	239	32	227	20	237	30	229	22	235	28	231	24	233	26
5	223	48	209	34	221	46	211	36	219	44	213	38	217	42	215	40
6	210	33	224	47	212	35	222	45	214	37	220	43	216	39	218	41
7	63	208	49	194	61	206	51	196	59	204	53	198	57	202	55	200
8	50	193	64	207	52	195	62	205	54	197	60	203	56	199	58	201
9	80	191	66	177	78	189	68	179	76	187	70	181	74	185	72	183
10	65	178	79	192	67	180	77	190	69	182	75	188	71	184	73	186
11	176	95	162	81	174	93	164	83	172	91	166	85	170	89	168	87
12	161	82	175	96	163	84	173	94	165	86	171	92	167	88	169	90
13	159	112	145	98	157	110	147	100	155	108	149	102	153	106	151	104
14	146	97	160	111	148	99	158	109	150	101	156	107	152	103	154	105
15	127	144	113	130	125	142	115	132	123	140	117	134	121	138	119	136
16	114	129	128	143	116	131	126	141	118	133	124	139	120	135	122	137

row numbers

2056 2056 2056 2056 2056 2056 2056 2056 2056

Figure 10.31 Franklin bent-right column.

Figure 10.32 Pattern of bent-right column.

column numbers

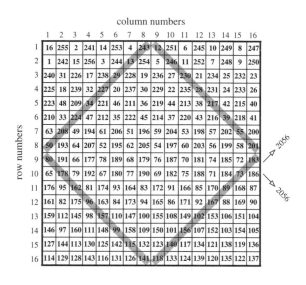

Figure 10.33   The magic diamond.

column numbers

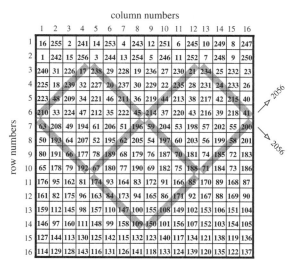

Figure 10.34   The diamond cluster.

146

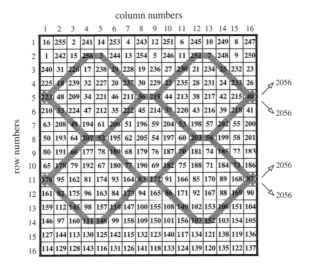

Figure 10.35 Five diamonds.

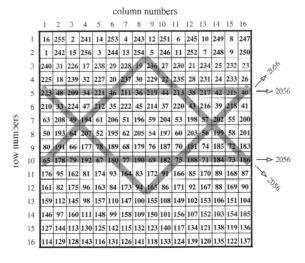

Figure 10.36 A magic Star of David.

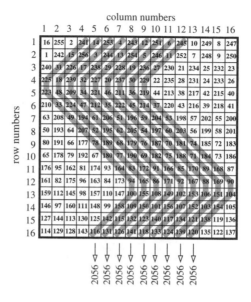

column numbers

	1	2	3	4	5	6	7	8	9	10	11	12	13	14	15	16
1	16	255	2	241	14	253	4	243	12	251	6	248	10	249	8	247
2	1	242	15	256	3	244	13	254	5	246	11	252	7	248	9	250
3	240	31	226	17	238	29	228	19	236	27	230	21	234	25	232	23
4	225	18	239	32	227	20	237	30	229	22	235	28	231	24	233	26
5	223	48	209	34	221	46	211	36	219	44	213	38	217	42	215	40
6	210	33	224	47	212	35	222	45	214	37	220	43	216	39	218	41
7	63	208	49	194	61	206	51	196	59	204	53	198	57	202	55	200
8	50	193	64	207	52	195	62	205	54	197	60	203	56	199	58	201
9	80	191	66	177	78	189	68	179	76	187	70	181	74	185	72	183
10	65	178	79	192	67	180	77	190	69	182	75	188	71	184	73	186
11	176	95	162	81	174	93	164	83	172	91	166	85	170	89	168	87
12	161	82	175	96	163	84	173	94	165	86	171	92	167	88	169	90
13	159	112	145	98	157	110	147	100	155	108	149	102	153	106	151	104
14	146	97	160	111	148	99	158	109	150	101	156	107	152	103	154	105
15	127	144	113	130	125	142	115	132	123	140	117	134	121	138	119	136
16	114	129	128	143	116	131	126	141	118	133	124	139	120	135	122	137

row numbers

2056 2056 2056 2056 2056 2056 2056 2056

Figure 10.37   Opposite left-right double-bent columns.

Figure 10.38   Pattern for opposite left-right double-bent columns.

148

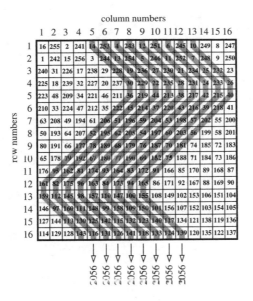

column numbers

	1	2	3	4	5	6	7	8	9	10	11	12	13	14	15	16
1	16	255	2	241	14	253	4	243	12	251	6	245	10	249	8	247
2	1	242	15	256	3	244	13	254	5	246	11	252	7	248	9	250
3	240	31	226	17	238	29	228	19	236	27	230	21	234	25	232	23
4	225	18	239	32	227	20	237	30	229	22	235	28	231	24	233	26
5	223	48	209	34	221	46	211	36	219	44	213	38	217	42	215	40
6	210	33	224	47	212	35	222	45	214	37	220	43	216	39	218	41
7	63	208	49	194	61	206	51	196	59	204	53	198	57	202	55	200
8	50	193	64	207	52	195	62	205	54	197	60	203	56	199	58	201
9	80	191	66	177	78	189	68	179	76	187	70	181	74	185	72	183
10	65	178	79	192	67	180	77	190	69	182	75	188	71	184	73	186
11	176	95	162	81	174	93	164	83	172	91	166	85	170	89	168	87
12	161	82	175	96	163	84	173	94	165	86	171	92	167	88	169	90
13	159	112	145	98	157	110	147	100	155	108	149	102	153	106	151	104
14	146	97	160	111	148	99	158	109	150	101	156	107	152	103	154	105
15	127	144	113	130	125	142	115	132	123	140	117	134	121	138	119	136
16	114	129	128	143	116	131	126	141	118	133	124	139	120	135	122	137

row numbers

2056 2056 2056 2056 2056 2056 2056 2056

Figure 10.39 Opposite right-left double-bent columns.

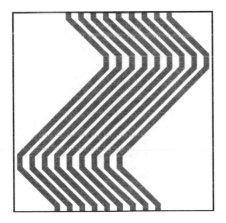

Figure 10.40 Pattern for opposite right-left double-bent columns.

149

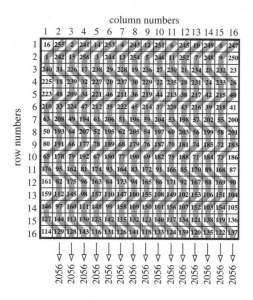

	1	2	3	4	5	6	7	8	9	10	11	12	13	14	15	16
1	16	255	2	241	14	253	4	243	12	251	6	245	10	249	8	247
2	1	242	15	256	3	244	13	254	5	246	11	252	7	248	9	250
3	240	31	226	17	238	29	228	19	236	27	230	21	234	25	232	23
4	225	18	239	32	227	20	237	30	229	22	235	28	231	24	233	26
5	223	48	209	34	221	46	211	36	219	44	213	38	217	42	215	40
6	210	33	224	47	212	35	222	45	214	37	220	43	216	39	218	41
7	63	208	49	194	61	206	51	196	59	204	53	198	57	202	55	200
8	50	193	64	207	52	195	62	205	54	197	60	203	56	199	58	201
9	80	191	66	177	78	189	68	179	76	187	70	181	74	185	72	183
10	65	178	79	192	67	180	77	190	69	182	75	188	71	184	73	186
11	176	95	162	81	174	93	164	83	172	91	166	85	170	89	168	87
12	161	82	175	96	163	84	173	94	165	86	171	92	167	88	169	90
13	159	112	145	98	157	110	147	100	155	108	149	102	153	106	151	104
14	146	97	160	111	148	99	158	109	150	101	156	107	152	103	154	105
15	127	144	113	130	125	142	115	132	123	140	117	134	121	138	119	136
16	114	129	128	143	116	131	126	141	118	133	124	139	120	135	122	137

column numbers

row numbers

2056 2056 2056 2056 2056 2056 2056 2056 2056 2056 2056 2056 2056 2056 2056 2056

Figure 10.41   Opposite left-right quadruple-bent columns.

Figure 10.42   Pattern for opposite left-right quadruple-bent columns.

150

column numbers

1 2 3 4 5 6 7 8 9 10 11 12 13 14 15 16

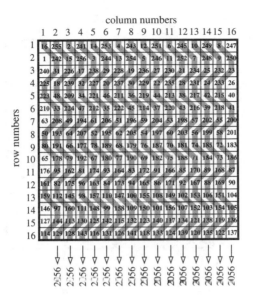

row numbers

	1	2	3	4	5	6	7	8	9	10	11	12	13	14	15	16
1	16	255	2	241	14	253	4	243	12	251	6	245	10	249	8	247
2	1	242	15	256	3	244	13	254	5	246	11	252	7	248	9	250
3	240	31	226	17	238	29	228	19	236	27	230	21	234	25	232	23
4	225	18	239	32	227	20	237	30	229	22	235	28	231	24	233	26
5	223	48	209	34	221	46	211	36	219	44	213	38	217	42	215	40
6	210	33	224	47	212	35	222	45	214	37	220	43	216	39	218	41
7	63	208	49	194	61	206	51	196	59	204	53	198	57	202	55	200
8	50	193	64	207	52	195	62	205	54	197	60	203	56	199	58	201
9	80	191	66	177	78	189	68	179	76	187	70	181	74	185	72	183
10	65	178	79	192	67	180	77	190	69	182	75	188	71	184	73	186
11	176	95	162	81	174	93	164	83	172	91	166	85	170	89	168	87
12	161	82	175	96	163	84	173	94	165	86	171	92	167	88	169	90
13	159	112	145	98	157	110	147	100	155	108	149	102	153	106	151	104
14	146	97	160	111	148	99	158	109	150	101	156	107	152	103	154	105
15	127	144	113	130	125	142	115	132	123	140	117	134	121	138	119	136
16	114	129	128	143	116	131	126	141	118	133	124	139	120	135	122	137

2056 2056 2056 2056 2056 2056 2056 2056 2056 2056 2056 2056 2056 2056 2056 2056

Figure 10.43 Opposite right-left quadruple-bent columns.

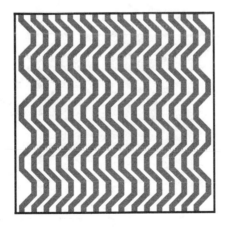

Figure 10.44 Pattern for opposite right-left quadruple-bent columns.

151

column numbers

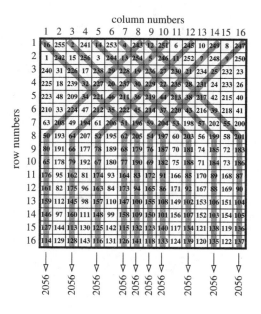

	1	2	3	4	5	6	7	8	9	10	11	12	13	14	15	16
1	16	255	2	241	14	253	4	243	12	251	6	248	10	249	8	247
2	1	242	15	256	3	244	13	254	5	246	11	252	7	248	9	250
3	240	31	226	17	238	29	228	19	236	27	230	21	234	25	232	23
4	225	18	239	32	227	20	237	30	229	22	235	28	231	24	233	26
5	223	48	209	34	221	46	211	36	219	44	213	38	217	42	215	40
6	210	33	224	47	212	35	222	45	214	37	220	43	216	39	218	41
7	63	208	49	194	61	206	51	196	59	204	53	198	57	202	55	200
8	50	193	64	207	52	195	62	205	54	197	60	203	56	199	58	201
9	80	191	66	177	78	189	68	179	76	187	70	181	74	185	72	183
10	65	178	79	192	67	180	77	190	69	182	75	188	71	184	73	186
11	176	95	162	81	174	93	164	83	172	91	166	85	170	89	168	87
12	161	82	175	96	163	84	173	94	165	86	171	92	167	88	169	90
13	159	112	145	98	157	110	147	100	155	108	149	102	153	106	151	104
14	146	97	160	111	148	99	158	109	150	101	156	107	152	103	154	105
15	127	144	113	130	125	142	115	132	123	140	117	134	121	138	119	136
16	114	129	128	143	116	131	126	141	118	133	124	139	120	135	122	137

row numbers

2056 2056 2056 2056 2056 2056 2056 2056 2056 2056

Figure 10.45 Down candelabras.

Figure 10.46 Pattern for down candelabras.

152

	1	2	3	4	5	6	7	8	9	10	11	12	13	14	15	16
1	16	255	2	241	14	253	4	243	12	251	6	245	10	249	8	247
2	1	242	15	256	3	244	13	254	5	246	11	252	7	248	9	250
3	240	31	226	17	238	29	228	19	236	27	230	21	234	25	232	23
4	225	18	239	32	227	20	237	30	229	22	235	28	231	24	233	26
5	223	48	209	34	221	46	211	36	219	44	213	38	217	42	215	40
6	210	33	224	47	212	35	222	45	214	37	220	43	216	39	218	41
7	63	208	49	194	61	206	51	196	59	204	53	198	57	202	55	200
8	50	193	64	207	52	195	62	205	54	197	60	203	56	199	58	201
9	80	191	66	177	78	189	68	179	76	187	70	181	74	185	72	183
10	65	178	79	192	67	180	77	190	69	182	75	188	71	184	73	186
11	176	95	162	81	174	93	164	83	172	91	166	85	170	89	168	87
12	161	82	175	96	163	84	173	94	165	86	171	92	167	88	169	90
13	159	112	145	98	157	110	147	100	155	108	149	102	153	106	151	104
14	146	97	160	111	148	99	158	109	150	101	156	107	152	103	154	105
15	127	144	113	130	125	142	115	132	123	140	117	134	121	138	119	136
16	114	129	128	143	116	131	126	141	118	133	124	139	120	135	122	137

row numbers

2056 2056 2056 2056 2056 205€ 205€ 2056 2056 2056

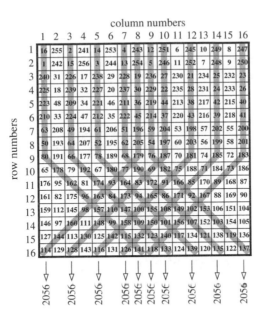

Figure 10.47 Up candelabras.

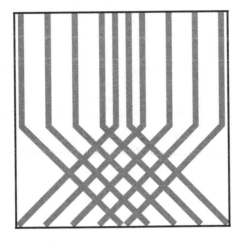

Figure 10.48 Pattern for up candelabras.

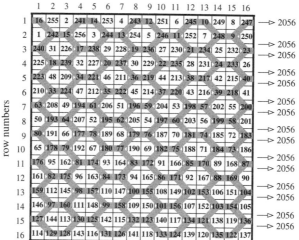

column numbers

	1	2	3	4	5	6	7	8	9	10	11	12	13	14	15	16	
1	16	255	2	241	14	253	4	243	12	251	6	245	10	249	8	247	⟶ 2056
2	1	242	15	256	3	244	13	254	5	246	11	252	7	248	9	250	⟶ 2056
3	240	31	226	17	238	29	228	19	236	27	230	21	234	25	232	23	⟶ 2056
4	225	18	239	32	227	20	237	30	229	22	235	28	231	24	233	26	⟶ 2056
5	223	48	209	34	221	46	211	36	219	44	213	38	217	42	215	40	⟶ 2056
6	210	33	224	47	212	35	222	45	214	37	220	43	216	39	218	41	⟶ 2056
7	63	208	49	194	61	206	51	196	59	204	53	198	57	202	55	200	⟶ 2056
8	50	193	64	207	52	195	62	205	54	197	60	203	56	199	58	201	⟶ 2056
9	80	191	66	177	78	189	68	179	76	187	70	181	74	185	72	183	⟶ 2056
10	65	178	79	192	67	180	77	190	69	182	75	188	71	184	73	186	⟶ 2056
11	176	95	162	81	174	93	164	83	172	91	166	85	170	89	168	87	⟶ 2056
12	161	82	175	96	163	84	173	94	165	86	171	92	167	88	169	90	⟶ 2056
13	159	112	145	98	157	110	147	100	155	108	149	102	153	106	151	104	⟶ 2056
14	146	97	160	111	148	99	158	109	150	101	156	107	152	103	154	105	⟶ 2056
15	127	144	113	130	125	142	115	132	123	140	117	134	121	138	119	136	⟶ 2056
16	114	129	128	143	116	131	126	141	118	133	124	139	120	135	122	137	⟶ 2056

row numbers

Figure 10.49 Rows in zig-zag.

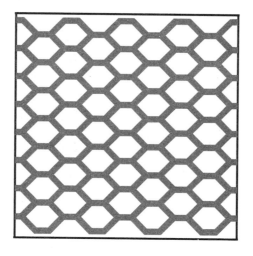

Figure 10.50 Pattern created with rows in zig-zag.

154

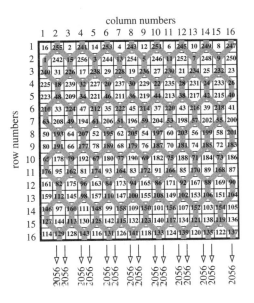

column numbers

	1	2	3	4	5	6	7	8	9	10	11	12	13	14	15	16
1	16	255	2	241	14	253	4	243	12	251	6	245	10	249	8	247
2	1	242	15	256	3	244	13	254	5	246	11	252	7	248	9	250
3	240	31	226	17	238	29	228	19	236	27	230	21	234	25	232	23
4	225	18	239	32	227	20	237	30	229	22	235	28	231	24	233	26
5	223	48	209	34	221	46	211	36	219	44	213	38	217	42	215	40
6	210	33	224	47	212	35	222	45	214	37	220	43	216	39	218	41
7	63	208	49	194	61	206	51	196	59	204	53	198	57	202	55	200
8	50	193	64	207	52	195	62	205	54	197	60	203	56	199	58	201
9	80	191	66	177	78	189	68	179	76	187	70	181	74	185	72	183
10	65	178	79	192	67	180	77	190	69	182	75	188	71	184	73	186
11	176	95	162	81	174	93	164	83	172	91	166	85	170	89	168	87
12	161	82	175	96	163	84	173	94	165	86	171	92	167	88	169	90
13	159	112	145	98	157	110	147	100	155	108	149	102	153	106	151	104
14	146	97	160	111	148	99	158	109	150	101	156	107	152	103	154	105
15	127	144	113	130	125	142	115	132	123	140	117	134	121	138	119	136
16	114	129	128	143	116	131	126	141	118	133	124	139	120	135	122	137

2056 2056 2056 2056 2056 2056 2056 2056 2056 2056 2056 2056 2056 2056 2056

Figure 10.51 Columns in zig-zag.

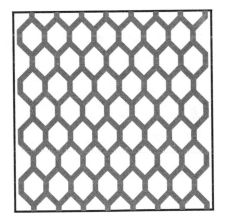

Figure 10.52 Pattern created with columns in zig-zag.

155

column numbers

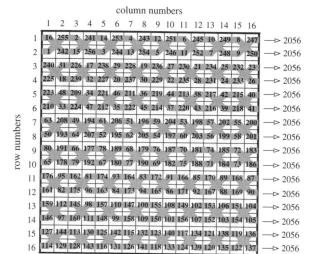

	1	2	3	4	5	6	7	8	9	10	11	12	13	14	15	16	
1	16	255	2	241	14	253	4	243	12	251	6	245	10	249	8	247	→ 2056
2	1	242	15	256	3	244	13	254	5	246	11	252	7	248	9	250	→ 2056
3	240	31	226	17	238	29	228	19	236	27	230	21	234	25	232	23	→ 2056
4	225	18	239	32	227	20	237	30	229	22	235	28	231	24	233	26	→ 2056
5	223	48	209	34	221	46	211	36	219	44	213	38	217	42	215	40	→ 2056
6	210	33	224	47	212	35	222	45	214	37	220	43	216	39	218	41	→ 2056
7	63	208	49	194	61	206	51	196	59	204	53	198	57	202	55	200	→ 2056
8	50	193	64	207	52	195	62	205	54	197	60	203	56	199	58	201	→ 2056
9	80	191	66	177	78	189	68	179	76	187	70	181	74	185	72	183	→ 2056
10	65	178	79	192	67	180	77	190	69	182	75	188	71	184	73	186	→ 2056
11	176	95	162	81	174	93	164	83	172	91	166	85	170	89	168	87	→ 2056
12	161	82	175	96	163	84	173	94	165	86	171	92	167	88	169	90	→ 2056
13	159	112	145	98	157	110	147	100	155	108	149	102	153	106	151	104	→ 2056
14	146	97	160	111	148	99	158	109	150	101	156	107	152	103	154	105	→ 2056
15	127	144	113	130	125	142	115	132	123	140	117	134	121	138	119	136	→ 2056
16	114	129	128	143	116	131	126	141	118	133	124	139	120	135	122	137	→ 2056

row numbers

Figure 10.53   Interlaced rows in zig-zag.

Figure 10.54   Pattern created with interlaced rows in zig-zag.

156

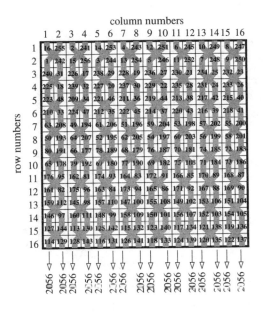

Figure 10.55 Interlaced columns in zig-zag.

Figure 10.56 Pattern created with interlaced columns in zig-zag.

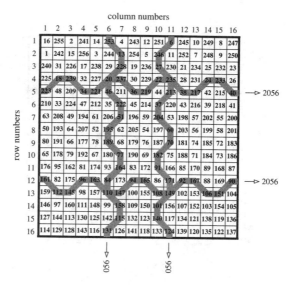

Figure 10.57 Ornament created with rows and columns in zig-zag.

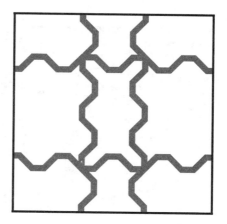

Figure 10.58 Pattern of ornament created with rows
and columns in zig-zag.

158

10.3   Magic Squares and Music

Sir Peter Maxwell Davies is considered one of the most important composers of the twentieth century (Figure 10.59). He has also been a noted conductor. He belongs to the British group interested in early twentieth-century music like that of Schoenberg, Webern, and Berg. Davies had a great interest in medieval history and music. John Michell's book *The View over Atlantis* [50], which delved into medieval and ancient lore, introduced Davies to magic squares, which he used to control melodic, harmonic, rhythmic, and proportional aspects of his compositions. He regarded the squares as being like prisms through which the basic material of a work (almost always a segment of a plainsong tune) is refracted. Davies employed the Ptolemaic planetary magic squares, that is, 9 for the Moon, 8 for Mercury, 7 for Venus, 6 for the Sun, 5 for Mars, 4 for Jupiter, and 3 for Saturn. He claimed he based each work on a specific square. His work *Image, Reflection, Shadow* is based on a plainsong tune *Lux Aeterna* (Light Eternal), and because light is related to the Sun, it required a magic square of order 6. His *Ave Maris Stella*, the first composition in which he used magic

Figure 10.59   Peter Maxwell Davies.

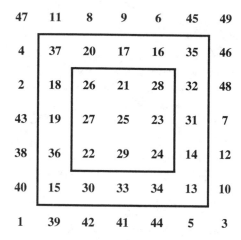

Figure 10.60   Nested magic square of orders 3, 5, and 7 that Davies
used in his Naxos Quartet.

squares (1975), referenced a connection between the Virgin and
the Moon, and therefore required a magic square of 9 [44].

In another composition, *Naxos Quartet No. 3*, honoring
St. Cecilia, the patron saint of music, and drawn from a plainsong hymn for St. Cecilia's Day, Davies uses an order 7 square in
which is nested a 5 square, which in turn has a 3 square nested in
it. (See Figure 10.60 for the magic square and Figure 10.61 for it
superimposed on the music [45]). (See Figure 8.13 for an order 12
magic square with four squares nested in it.) In 1988 Davies' *Violin
Concerto* was to be performed by Isaac Stern and the Los Ange-
les Philharmonic Orchestra, and Davies was being interviewed in
connection to the performance [79]. The interviewer eventually
got around to the subject of magic squares.

The interviewer asked You have used a technique using magic
squares. Can you shed a little bit of light on that?

Davies replied If you use those traditional magic squares—and
I'm thinking of one that I've just been using—and you use
that consistently, it gives certain symmetries because number

patterns recur. A phrase will balance itself. The notes will recur at a transition, and it's a very useful basis, a framework to hang one's ideas. One thing I would like to point out though is that I don't sit there with a chart of paper in front of me working music with that. You learn the thing so that you can carry it around in your head; otherwise it would be tedious. It's really an aid to composition and an aid in one's structure work and symmetries—both rhythmically and as far as the intervals go. It's not for the ornamental.

Interviewer One more question about magic squares. Does it bring a little mysticism to your music or is it purely a mathematical device?

Davies I think it's probably both. But I think too, that one has to remember that numbers—one, two, three, four and so on up to probably nine, but certainly the first four—when they were originally used, I doubt whether they were used so much for counting as for their quality. I mean their almost religious quality. To actually count is already a magic thing. And numbers and certainly magic squares, they have got enormous magic quality and I think I'm very conscious of the magic of numbers. Although this does tend to become debased when one counts one's change at the supermarket, but when you actually begin to think about the primary numbers and their application in computer techniques, then it again becomes magic.

Other Composers

Other twentieth-century composers who employed magic squares in their works included Italian Bruno Maderna, who made use of the mathematical ordering of magic squares for the distribution of timbre and the unfolding of sound events in his compositions of the 1950s and '60s. French composer Marcelle Manziarly demonstrated an interest in precompositional structure in her piano piece *Stances* (1964), which is based on a magic square. Hungarian

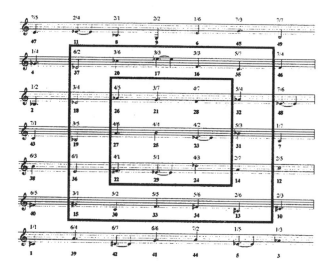

Figure 10.61   Nested magic square of order 3, 5, and 7
superimposed on the music.

Laszlo Dubrovay titled his violin composition *Magic Square* (1975). That was a good year for magic squares in music because the same year English composer Malcolm Singer titled his piece *Triflutes in Saturn's Magic Square*. Later in the twentieth century, Irish composer-producer David Byers wrote compositions that combined strong linearity with intricate polyphony. He structured pitch material through magic squares and note rows of varying lengths—devices that he called nets and sieves [59].

Magic Squares and Indian Drums

In music in another part of the world, magic squares have also made their contribution. This application of magic squares is in the area of rhythm, to help keep the beat. It is peculiar to music of India, where magic squares have been known since ancient times. Information on this subject comes to us from Jonathon Dimond, an

Figure 10.62   Indian playing the tabla.

Australian who obtained degrees from the Queensland Conservatory of Music, and the New England Conservatory in Boston, and has undergone vigorous training in north Indian classical music in Pune, India.

The two major characteristics of Indian classical music are raga, the structure of melody, and tala, the structure of rhythm. The melody deals with the rise and fall of sounds, and the tala concerns the time beats of ragas. The principal instrument for performing the raga is a string instrument, the sitar, whereas that for playing the tala is called the tabla, the name for a pair of different-size hand drums, each containing a round black spot (Figure 10.62), which gives a different pitch from the white part of the goat skin surface. This gives the player the opportunity for improvisation. Tabla are played in a sitting position.

Dimond believes that the unity and logic of magic squares attracted Indian musicians to applying them to their music, and he says they seem perfect for applying to time cycles and additive rhythm. For example, see the following 3 × 3 modified magic square [81] (Figure 10.63).

Figure 10.63   Modified order 3 square.

Obviously, this is not a true magic square. Only the first row and the first column add to 15, but even here there is a repetition of numbers not permitted in traditional magic squares. In a traditional magic square with nine cells only numbers 1 to 9 are used, and each row and each column must add to give the magic constant of 15. This square has been modified so that the same relationship of the middle number to the total sum of the numbers of the magic square applies to the new purpose. That is, the sum of all the numbers in the magic square equals the central number times nine. The size of the magic square, indicated by this product, tells how many pulses or subdivisions there are in the whole sequence. This gives us ideas as to the meter, where to start and to end. In the example, $8 \times 9 = 72$ represents the magic square's size. It could be applied to a passage lasting 18 beats in sixteenth notes: $18 \times 4 = 72$.

The idea of the magic square as modified is to give patterns for the rhythm. Note that there is a sequence of a difference of two between the numbers in the first row, a difference of three in the second row, and a difference of four in the third row. Another pattern is found in the sum of each row, where there is an increasing sum of nine between rows, namely, 15, 24, and 33. According to Dimond, the three-part structure of the 3×3 square makes for a great rhythm sequence, which naturally gravitates toward the next beat after the end of each row. With regard to the specific internal groupings that make the pattern more audible, the following is based on the numbers in the rows of the square, where the number in each cell represents the number of beats, and the first number in the row specifies the number of accents or *"hits"* in that row. For the first row, the suggested groupings are:

$$3 = 1 + 1 + 1,$$
$$5 = 2 + 2 + 1,$$
$$7 = 3 + 2 + 2.$$

For the 3, each 1 has an equal hit. For the 5, the accent is put on the first beat of each of the 2s and on the 1. For the 7, the first of the three beats is accented and the first of each of the two beats. It is clear that the first beat in each grouping is accented; in this row giving three hits in each case. For the second row the suggested groupings are:

$$5 \quad 1 + 1 + 1 + 1 + 1,$$
$$8 = 3 + 2 + 1 + 1 + 1,$$
$$11 = 4 + 4 + 1 + 1 + 1.$$

In the second row, the five 1s all have equal accents. With the 8, the first beat of the 3, the first beat of the 2, and the three 1s are all accented, giving the five hits. In case of the 11, the first beat of each 4 and the three 1s are accented, giving five hits, as with the 5 and 8. The third row has the following groupings:

$$7 = 1 + 1 + 1 + 1 + 1 + 1 + 1,$$
$$11 = 3 + 3 + 1 + 1 + 1 + 1 + 1,$$
$$15 = 5 + 5 + 1 + 1 + 1 + 1 + 1.$$

Here the seven 1s have equal hits. With the 11, each 3 and the five 1s receive accents, and with the 15, each 5 and the five 1s are accented, giving a total of seven hits, or attacks, as they are sometimes called.

3	5	7 (12)
5	8	11 (12)
7	11	15

Figure 10.64   Modified order 3 square with gaps of 12.

But this size 72 magic square doesn't end with passages lasting 18 beats in sixteenth notes. There are rests or gaps inserted between the first and second and the second and third rows, which can add to the building feeling of this cadence. Dimond says that by choosing different values for the gap you can adapt one magic square into many musical situations. An example is the same magic square with two gaps of 12 (Figure 10.64).

Here the size of the magic square is $72 + (2 \times 12) = 96$, and a magic square of 96 would fill 12 bars of 4/4 in eighth notes, or 8 bars of 4/4 in eighth-note triplets. The first note of each of the gaps is accented, which makes the first note immediately following the magic square seem like the third big attack of the series. In other words, it will strongly feel like the destination of the cadence.

Dimond also introduces other squares of different sizes, where $9 \times 9 = 81$ and $7 \times 9 = 63$. For further information the reader will be interested in an article, "An Introduction to Tabla: North Indian Rhythm and its Applications," by Jonathan Dimond [82].

Dimond refers us to a piano composition by Dmitri N. Smirnov titled *Two Magic Squares* [109]. The composer said:

> This is the 1st piece from TWO MAGIC SQUARES. It is based on a 12-tone row (c-f-b-f#-c#-g-d-g#-d#-a#-e-a) explores the three-celled magic square known from ancient Chinese and Arabian manuscripts. It consists of nine numbers, every three of which (verticals, horizontals and diagonals) have in sum 15. I took this idea to organize the

rhythmic structure of the piece: every bar of the music corresponds to one of these numbers, having the same quantity of the rhythmic units, so in every three bars there are 15 rhythmic units. The process of exhaustion of all variants has 16 steps: including horizontals, verticals and diagonals in all directions.

You can listen to this piece via the Internet (it is just two minutes long) and decide for yourself whether you find it to be a musical magic square. Dimond reviewed it and offered his opinion:

> As somebody interested in Magic Squares myself, and their application to music, I was curious to listen to this composition to hear what it sounded like. Though very different in style to my Indian-Jazz fusion approach, this Russian composer has brought to the piano an accessible (dare I say "jazzy") approach to the mathematics, which sound serial-esque but bears the composer's obviously musical ear.

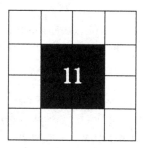

The Fourth Dimension

11.1 Magic Square in Three Dimensions

Figure 11.1 shows a magic square in three dimensions, a sculpture by Patric Ireland in Eaton Fine Art Gallery, West Palm Beach, Florida [100]. It is a 3 × 3 square. The numbers are indicated by counting the numbers of blocks in each column and dividing by two.

Figure 11.1   Magic square in three dimensions.

11.2   Leaping into the Higher Dimensions

Hold your breath! Here we go, leaping into the fourth and higher dimensions of space. But don't worry, we will approach them gradually, one dimension at a time.

In early times, artists showed three-dimensional space in two dimensions, but their paintings looked flat. Following the invention of perspective by Brunelleschi and its publication in the book *On Painting* by Alberti in 1435, the technique was acquired by artists and allowed them to give their work the appearance of depth while still painting on flat, two-dimensional surfaces (Figure 11.2). Although we cannot visualize in our minds anything beyond three dimensions, mathematicians have been able to work in higher dimensions, and Einstein in his theory of relativity, referred to time as the fourth dimension.

Though we cannot actually visualize the fourth dimension, we can use an analogy to help us to try do so. Imagine that we are

Figure 11.2   Leonardo da Vinci (1452–1519) used perspective to
portray a three-dimensional object in two-dimensional space.

trying to get in touch with a friend who lives in New York City.
We know he lives on Broadway and 110th Street in apartment
16B on the sixteenth floor. We have the length, width, and height
dimensions to find him. But he is only there from 5 to 6 P.M. Now
that we have this fourth dimension, the time dimension, we can
find him. Time may not be the fourth space dimension, as we shall
see later in this chapter, but at least we can impress our friends by
agreeing with Einstein.

In England in 1884, the Anglican clergyman, teacher, and
mathematician Edwin A. Abbott wrote *Flatland* [1], a social satire
on the stratified Victorian society. It is a classic that is still in print
and has been the subject of many books, movies, games, and even
an opera. Abbott's allegorical Flatland is a two-dimensional world
whose citizens are triangles, squares, polygons, and circles living in
a planar landscape with no ups or downs.

In Flatland, social standing depends on your number of sides: triangles with three sides are the laborers and soldiers, on the bottom of the pack, whereas circles, with an infinite number of sides, the priests, are on the top. In between are the middle class, squares, with the nobility being polygons with more than four sides. The women are different; they have no sides. They are just straight lines pointed at the ends, like a needle with both ends pointed. They do not fit the status category, and the author says "a female in Flatland is a creature by no means to be trifled with."

The story is told by the principal character, A Square, a respectable member of the middle class. One day a Sphere comes to Flatland, appearing in the guise of a circle, and takes A Square with him into Spaceland, the outer world of three dimensions. Of course, A Square is flabbergasted at everything he sees. Objects rising and falling are the most perplexing to him: there is no gravity in Flatland. When he goes back to Flatland and relates what he has seen in Spaceland nobody believes him. They ridicule him and he eventually lands in jail as a troublemaker.

After he discovered Spaceland with its three dimensions, he began to wonder if so whether there could be three dimensions, and if so why not four, five, or six dimensions? The Sphere considered this ridiculous and scolded him for such foolish speculation.

Like A Square with his vision, we plan to make the fourth dimension more visible and understandable by the same process of analogy that guided him. We shall proceed in this chapter up the dimensional stairs until we have reached the fourth flight.

A *hypercube* is defined as a cube in a dimension higher than three. The word *tesseract* refers specifically to a cube in the fourth dimension. We owe this word to a mathematician and science fiction writer, Charles Howard Hinton (1853–1907), who wrote about the fourth dimension [99]. Hinton received an MA from Oxford and taught at Princeton and the University of Minnesota. He invented a baseball pitching machine, capable of variable speeds and throwing curveballs, that was used by the teams at these colleges. It had one unusual property: it was powered by gunpowder and caused several injuries, resulting in Hinton being dismissed from Princeton.

Although the tesseract owes its name to one mathematician, another brought it and the hypercube into the world of magic squares. The second mathematician was John R. Hendricks, a Canadian meteorologist for whom magic squares were a lifelong hobby. His contributions to the field of magic squares have been the most creative of the twentieth century. Along with many other things, he produced and published four-, five-, and six-dimensional hypercubes, as well as inlaid magic cubes and hypercubes. He has written prolifically in the *Journal of Recreational Mathematics* and other mathematical journals, in addition to lecturing and teaching on the subject of magic squares [80].

Though it is true that we cannot visualize the fourth and higher dimensions, we can by logical analogy understand these higher dimensions as we understand lower dimensions. Our eyes see what the camera sees when it produces a two-dimensional photograph. Although our eyes see only two dimensions, our brains supply the perspective to give us the illusion that we see three dimensions. This works for us as reality. Thus, in the following pages we employ logical analogy with diagrams to advance from zero dimension to the fourth dimension. We produce the projections of these dimensions on paper, that is, on two-dimensional space, with the illusion of three- and four-dimensional space.

11.3   Zero-Dimensional Space

A *point* is is said to be an object in *zero-dimensional space*. We cannot really see a point because it has no dimension, but it can be represented, for example, by a bullet as A in Figure 11.3(a).

The projection of a point over a line, Figure 11.3(b), or over a plane, Figure 11.3(c), will always be a point. Figure 11.3(b) shows a

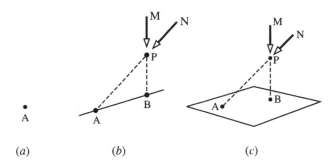

Figure 11.3   (a) Zero-dimensional space; (b) Projection of a point on a line (one-dimensional space); (c) Projection of a point on a plane (two-dimensional space).

point P and its projection B along the direction M and its projection A along the direction N. Similarly for the projections on a plane shown in Figure 11.3(c).

11.4   One-Dimensional Space

The displacement of zero-dimensional space generates *one-dimensional space*. The displacement of zero-dimensional space, represented by the point Q in Figure 11.4(a), to the point P generates one-dimensional space. The figure obtained is called a *line segment*. The trajectory of the point Q can be a straight line, as shown in Figure 11.4(a), or any curved line.

Figure 11.4(b) shows AC, the projection of the line segment QP, which is in the one-dimensional space, on the two-dimensional space represented by the plane.

Depending on the direction of the projection, the figure on the plane will be a line segment or a point. A light beam along the direction M will project the line segment QP as the point B on the

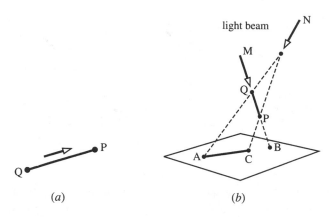

Figure 11.4 (a) Generation of one-dimensional space; (b) Projection of the one-dimensional object QP on two-dimensional space.

plane, which is illustrated in Figure 11.4. A light beam along the direction N will project the line segment QP as the line segment AC on the plane. The point B or the line segment AC are exactly the images that an observer will see if looking along the directions M or N, respectively.

The line segment AC is called the projection, along the direction N, of a one-dimensional object, represented by the line segment QP, on a two-dimensional space, represented by the plane. Similarly, the point B is the projection of the line segment QP along another direction, namely, M. Therefore, different directions of projection of a one-dimensional space object will give different images on two-dimensional space.

11.5   Two-Dimensional Space

The displacement of one-dimensional space generates *two-dimensional space*. Figure 11.5(a) shows the line segment AB displacing a distance equal to the length $AD = BC$ along a

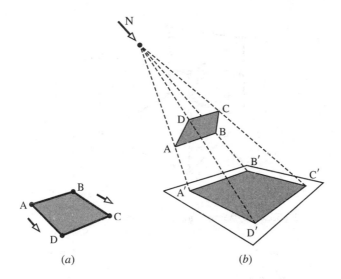

Figure 11.5   (a) Generation of a two-dimensional space; (b) Projection of the two-dimensional object ABCD on two-dimensional space.

perpendicular direction indicated by the arrows. This displacement generates the square ABCD, which is a two-dimensional figure.

Figure 11.5(b) shows the projection of a two-dimensional object, a square, on two-dimensional space, represented by the plane. The square ABCD is projected on the plane as the figure A' B' C' D'. If a light beam has the direction N, then the figure A' B' C' D' will be the in shade due to the square ABCD.

11.6   Three-Dimensional Space

The displacement of a two-dimensional space generates a *three-dimensional space*. Figure 11.6 shows an example where the square ABCD, which is a two-dimensional space object, is displaced until

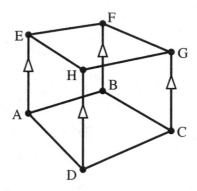

Figure 11.6   Generation of three-dimensional space.

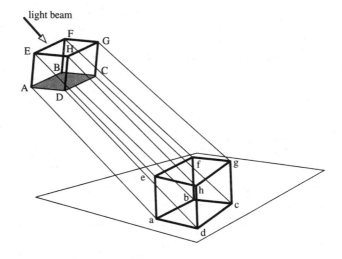

Figure 11.7   Projection of a cube on a plane.

it reaches the position represented by the square *EFGH*. The line segments *AE, BF, CG,* and *DH* complete the cubic shape, which is a three-dimensional object.

Figure 11.7 shows one of the possible projections of the cube *ABCD-EFGH*, which is a three-dimensional object, on a plane,

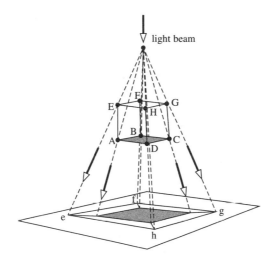

Figure 11.8   Projection of a cube on a plane along a direction
perpendicular to two opposite faces.

that is, a two-dimensional space. The projection contains two
squares ABCD and efgh, which are the projections of the lower
and upper surfaces of the cube, respectively. The projection of the
vertical lines AE, BF, CG, and DH of the cube are represented on
the plane by lines connecting the points Ee, Ff, Gg, and Hh of the
two squares.

Figure 11.8 shows the projection of a cube by a light beam at
right angles with respect to two opposite faces of the cube on a
plane, which is parallel to those surfaces of the cube. The projection of the upper surface EFGH is represented on the plane by the
square efgh. The projection of the lower surface ABCD gives the
gray square on the plane. Therefore, a projection due to a light
beam perpendicular to a face of the cube will be represented on the
plane by one square inside the other.

The cubes in Figures 11.7 and 11.8 have different projections
due to different directions of the beam of light.

11.7   Four-Dimensional Space and Beyond

The displacement of three-dimensional space generates *four-dimensional space*. Figure 11.6 shows that the displacement on a plane of the projection of a square from the position *ABCD* to the position *EFGH* generates the projection of a cube on the same plane, which is given by *ABCD-EFGH*. By the same technique, if the projection *ABCD-EFGH* of a cube on a plane is displaced to the position *A′ B′ C′ D′–E′ F′ G′ H′* it generates one of the infinite possible projections of a four-dimensional object called a tesseract, which is shown in Figure 11.9.

Figure 11.8 shows that the projection of a cube at right angles to two opposite faces, on a plane parallel to those faces, gives a square (represented in gray) inside another square.

Similarly, Figure 11.10 shows that the projection, at right angles to two opposite faces of a tesseract, on a plane parallel to those faces gives the projection of a cube (represented in gray) inside another cube.

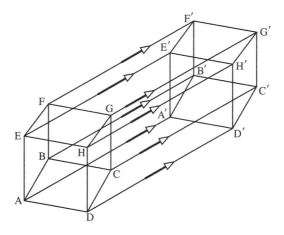

Figure 11.9 Projection of a tesseract on two-dimensional space, which is represented by the plane of the sheet of the paper.

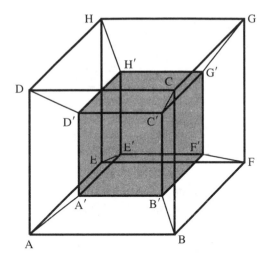

Figure 11.10 Projection of a tesseract along a direction at right angles to one of its faces on the plane of the sheet of paper.

Projections on a Two-Dimensional Plane

The projection on a two-dimensional plane is shown in Figures 11.11 and 11.12.

Projection of Each Cubic Face of the Tesseract

The projections of the eight cubes making a tesseract are better identified using the projection in Figure 11.13.

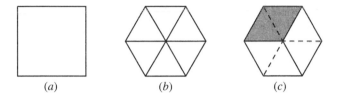

| (a) | (b) | (c) |

Figure 11.11 (a) Projection of a square on a two-dimensional plane; (b) Projection of a cube, along its space-diagonal, on a two-dimensional plane; (c) The cube in (b) shaded and in perspective.

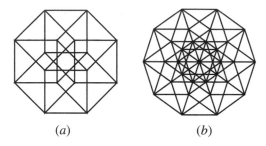

(*a*)   (*b*)

Figure 11.12 (*a*) Projection of a tesseract, along its space-diagonal, on a two-dimensional plane; (*b*) Projection of a fifth-dimension hypercube on a two-dimensional plane.

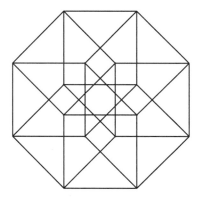

Figure 11.13 Projection of a tesseract on a plane.

Do you see the eight cubes in the tesseract in Figure 11.13? No? Examine Figures 11.14 to 11.17 and see if a little shading helps.

Figures 11.14 to 11.17 show, in gray, the projection of two faces. Because each face is a cube, the projection of the face of a tesseract from the fourth-dimensional space into a two-dimensional space is similar to the projection of a cube from the three-dimensional space onto the two-dimensional space.

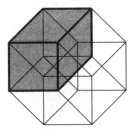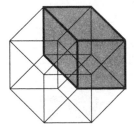

Figure 11.14   Projection of a tesseract showing the projection
of two of its cubic faces.

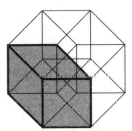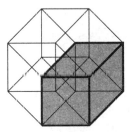

Figure 11.15   Projection of a tesseract showing the projection
of two of its cubic faces.

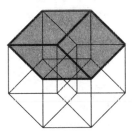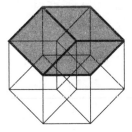

Figure 11.16   Projection of a tesseract showing the projection
of two of its cubic faces.

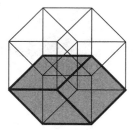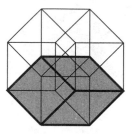

Figure 11.17   Projection of a tesseract showing the projection
of two of its cubic faces.

Magic Tesseract

The tesseract is the four-dimensional analog of the three-dimensional cube. In mathematics a tesseract is described as a regular convex 4-polytope whose boundary consists of eight cubical cells. (A 4-polytope is a closed four-dimensional many-sided object, sometimes called a polychron, with vertices, edges, faces, and cells.) The tesseract is composed of 8 cubic faces, called facets, 16 vertices, 32 edges, and 24 squares.

A magic tesseract contains the numbers $1, 2, \ldots, n^4$ where n is the order of the tesseract. Its magic constant is given by the expression

$$M_n = \frac{n}{2}(n^4 + 1). \tag{11.1}$$

Figure 11.18 shows the projection of the eight cubes that make a tesseract. The direction of projection was chosen such that two of the eight cubic faces have the well-known projection of a cube. The other six cubes are connecting each face of the front cube to the correspondent face of the back cube.

The third-order magic tesseract shown in Figure 11.18 was created by Hendricks [26].

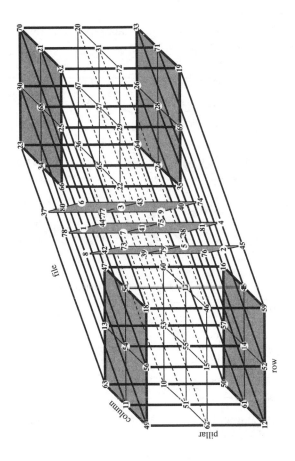

Figure 11.18 Projection of a tesseract on a two-dimensional plane.

As a $3 \times 3 \times 3$ standard tesseract it has $n^4 = 3^4 = 81$ numbers $1, 2, \ldots, 81$. Its magic constant is evaluated as

$$M_3 = \frac{n}{2}(n^4 + 1) = \frac{3}{2}(3^4 + 1) = \frac{3}{2}(81 + 1) = 123; \quad (11.2)$$

the addition of each number along any edge equals the magic constant.

The addition of each number along the rows on the lower square on the frontal cube writes

$$M_3 = 12 + 61 + 50 = 123,$$
$$M_3 = 52 + 14 + 57 = 123,$$
$$M_3 = 59 + 48 + 16 = 123. \tag{11.3}$$

The addition of each number along the columns on the middle square on the frontal cube writes

$$M_3 = 62 + 15 + 46 = 123,$$
$$M_3 = 51 + 55 + 17 = 123,$$
$$M_3 = 10 + 53 + 60 = 123. \tag{11.4}$$

The addition of each number along the pillars on the left square on the frontal cube writes

$$M_3 = 12 + 62 + 49 = 123,$$
$$M_3 = 61 + 51 + 11 = 123,$$
$$M_3 = 50 + 10 + 63 = 123. \tag{11.5}$$

The addition of each number along the files connecting the right side of the lower squares on the front and back cubes writes

$$M_3 = 59 + 45 + 19 = 123,$$
$$M_3 = 48 + 4 + 71 = 123,$$
$$M_3 = 16 + 74 + 33 = 123. \tag{11.6}$$

Projection of Each Face of the Tesseract

Figure 11.19 shows a projection of a tesseract on a two-dimensional plane: the sheet of paper on which it is drawn. The projection of two of its cubic faces are shown as a projection of cubes in wireframe representation. Shown in gray, in Figure 11.19(a), is the projection of another cubic face of the tesseract, which is connected to the two in wireframe projection.

Figures 11.19(b), 11.20(a) and (b), 11.21(a) and (b), and 11.22(a) and (b) show the detail of each side of the gray cube and the numbers of the third-order magic tesseact. The addition of each number on each edge equals the magic constant 130.

Because the tesseract has four dimensions, the fourth dimension, which is orthogonal to the other three, has been given the name *file*. See Figures 11.19 to 11.22.

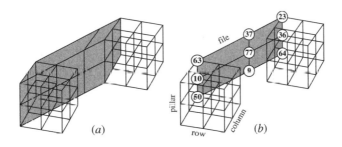

Figure 11.19   (*a*) Projection of a tesseract on two-dimensional space represented by a plane showing, in gray, one of its cubic facets; (*b*) The projection of the back side.

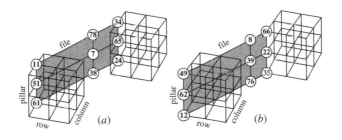

Figure 11.20   (*a*) Projection of the middle side; (*b*) Projection of the front side.

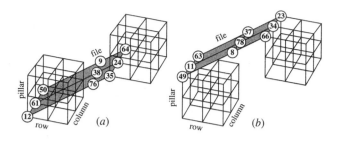

Figure 11.21   (a) Projection of the lower base; (b) Projection of the upper base.

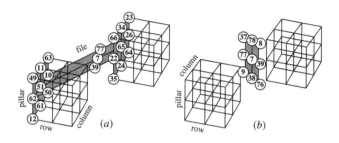

Figure 11.22   (a) Projection of the middle base; (b) Projection of the middle side.

11.8   Final Comments on Tesseracts

A third-order tesseract in the fourth dimension is as much as we can swallow in this book, but it may be of interest that magic tesseracts up to the eighth dimension have been constructed by Daniel M. Collinson and tesseracts up to order 9 by Meredith Houlton using John Hendricks's methods. Hendricks created the first perfect magic tesseract, an order 16, which is the smallest one possible. He uses numbers from 1 to 65,536 and has a magic constant of 524,296. It has been shown that there are only 58 magic tesseracts of order 3, and all have been found and published [98].

Practical Applications
of Magic Squares

12.1   Magic Squares and Statistics

A practical use of magic squares is in statistics. Statistics is a technical, mathematical subject practiced by specialists but is very
important to everyone. Thanks to statistics, we have an estimate
of how long we will live, the chance of being struck by lightning or
a comet, and whether our favorite horse will win the Kentucky
Derby. Nothing in life is certain, but with statistical methods
developed in the last century, we can make useful estimates of
the possibilities of these occurrences. These methods have become

important in the insurance business, finance, agriculture, weather forecasting, sports and gambling, and scientific and social research.

The number of births, deaths, marriages, divorces, as well as business records and weather data, are all statistics. *Statistics* refer to numerical data as well as the collection, organization, and analysis of these data. The word *statistic* comes from the Italian word *statista*, meaning statesman, because the first statisticians were government officials who collected such information. Official government statistics are as old as recorded history. Emperor Yao took a census of the population of China in the year 2238 B.C. About 1086 A.D., William the Conqueror ordered the writing of the *Domesday Book*, a record of the ownership and value of the lands in England, which was England's first statistical abstract. Because of Henry VII's fear of the plague, England began to register its dead in 1532. About this time, French law required the clergy to register baptisms, deaths, and marriages. In our industrial, scientific age, statistics is practiced by mathematicians who deal with masses of numerical data. That is where magic squares and Latin squares come in.

J. P. N. Phillips, of the Institute of Psychiatry of the Maudsley Hospital in London, described the use of magic squares in statistical methods of analyzing experiments (Phillips [55]). More applications, however, have been with the related Latin squares.

Latin squares are an offshoot of magic squares created in the eighteenth century by Swiss mathematician L. Euler, who called them "a new kind of magic squares." He used Latin letters to fill the cells, instead of Greek letters or numbers, so they have been called Latin squares. Today, numbers are usually used especially for larger order Latin squares. In Latin squares the cells are filled in a way so that each letter or number occurs only once in every row and once in every column. They have the same square design as magic squares, and the sum of each row and column always gives the magic constant. The diagonals of Latin squares do not necessarily give the magic sum, and Latin squares are nontraditional in other ways as well. In the traditional magic square of order n, the square

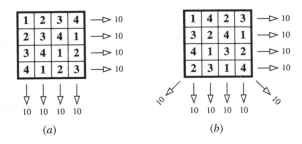

Figure 12.1   Latin squares. (a) Semi-perfect magic square;
(b) Perfect magic square.

must contain all the numbers from 1 to n^2. For example, a square
of order 8 must use all the numbers from 1 to 64 (8^2). This is
not required in Latin squares, and in addition, they may use zero, a
concept that was not known or recognized by early mathematicians
who devised magic squares.

In Figure 12.1(a) and (b) are shown two of the possible
Latin squares of order 4. Note that that the square shown in
Figure 12.1(b) has rows, columns, and diagonals that give the magic
constant.

These Latin squares begin with 1 but don't go to 16 (4^2), and
they may repeat the same numbers in every row and column, but
not within each row or column. Each row and each column always
adds up to the same sum. In their book *Latin Squares and Their
Applications*, Denes and Keedwell, in the chapter "Connections
between Latin Squares and Magic Squares" [13], illustrate squares
like those in Figure 12.1 going up as high as order 21 [13, p. 93].
Denes and Keedwell also illustrate a Latin square of order 9 that
starts with zero and goes as high as 88 [13, p. 211]. Obviously, with
order 9 (n^2 is 81) and so on, while there may be no repeat numbers,
some numbers are skipped (in this case all ending with 9, that is,
09, 19, 29, etc.). This is also permissible in Latin squares. It should
be stated that these deviations from traditional magic squares are
also freely accepted and practiced by modern puzzle enthusiasts

who create magic squares. As was noted at the beginning of this book, the Sudoku puzzle is a modern creation based on the 9 × 9 Latin square.

In the early 1920s Ronald Fisher of the University of Cambridge, one of the world's leading statisticians, showed how Latin squares could be used to make experiments more efficient, saving time and money; they are now employed in setting up experiments and analyzing complex data. In a chapter devoted to the practical application of Latin squares, Denes and Keedwell [13, p. 348] refer to their application in statistical design. A good example is found in a book titled *An Introduction to Agricultural Statistics* [10]. The author considers a feeding experiment in which animals of equal weight are tested with regard to sex (heifer, bull, and steer), genetic influence of Brahma in-breeding, and quantity of feed. The selected animals had either 25, or greater than 50 percent Brahma blood. Sex was interpreted in the rows, genetic influence in the columns, and feed quantities as percentage of body weight. Each sex type and each Brahma type animal was given an amount of feed equal to 1.5 percent, 2.5 percent, or 3.5 percent of its body weight, 1, 2, 3 in Table 12.1. The setup was for a multiple of three animals of each type of equal weight at the beginning of the experiment. After 120 days, the weight of all the animals was determined. Table 12.1 is the experimental layout with a Latin square of order 3, which was employed for the statistical calculation of the experimental results.

TABLE 12.1 Latin square layout for animal experiments.

Sex	% Brahma Blood		
	0	25	50+
Heifer	2	1	3
Bull	3	2	1
Steer	1	3	2

The total weight in pounds for group 1 was 112, for group 2 was 454, and for group 3 was at 832 in Table 12.2. The average weight per animal (the above weights divided by 3) for group 1 was 37.3 pounds, for group 2 was 151.3 pounds, and for group 3 was 277 pounds. Following the statistical calculations in the book [10, pp. 140–141], the author states that the average daily gain is not the same for the three feed rations, thus the differences are significant. Simple observation of the numbers in the three cases would lead the observer to conclude that greater amount of feed was worthwhile because it produced the best results. But how significant were the differences? The statistical analysis tells us that these differences due to chance alone are less than 1 in a 1,000, or putting it another way, in 999 cases out of 1,000, the differences were due to the amount of feed that was given.

However, if you wish to know specifically how much difference in weight between the animals would be required to make a significant difference, you would require additional calculations. These were run by Dr. Mark Yang in the Statistics Department at the University of Florida. They show that a difference of 10.7 pounds would be necessary for a 95 percent confidence level, that is, in only 5 percent of the time (1 case in 20) would the difference be due to just random sampling, experimental error, and so on. If that degree of confidence would not satisfy you, then further

TABLE 12.2 Experimental results with the weight increase of the animals, in parentheses, after 120 days.

Sex	% Brahma Blood		
	0	25	50+
Heifer	2 (150)	1 (28)	3 (256)
Bull	3 (288)	2 (146)	1 (24)
Steer	1 (60)	3 (288)	2 (156)

calculations show that a difference of 78.9 pounds would give a 99.9 percent confidence level, or odds of only 1 in a 1,000 chance that the difference could be due to chance alone. As can be seen from the experimental results, the differences in average weights exceed 78.9 pounds. So we may be quite confident of the conclusion that the extra feed was indeed responsible for the weight gain, regardless of sex or Brahma type.

Another statistical experiment with Latin squares is presented in Patterson's *Statistical Technique in Agricultural Research* [53]. The experiment in this case is similar to the former because it relates to the amount of fertilizer to be applied to a plot of land for its effect on the yield of sugarcane. Latin square design can be used in business experiments as well, as in a case of consumer studies of grocery stores. With stores in a large city, people shop differently on different days and generally shop in stores near their neighborhood. Because neighborhoods are related to income level, that serves as an indication of consumer income. In the Latin square design, rows could be different locations and columns could be days of the week. The treatment could be the price of watermelons, and the number sold at each price could be the data collected. In medical research, Latin squares have compared the effect of different drugs, and in industry, for example, have compared the effect of different types of concrete in a construction project.

12.2   Error Correcting Codes

Another important practical application of Latin squares has been in their use in producing error correction codes. Error correction is essential in modern electronic communication. This includes wireless network, satellite radio, digital TV, cellular telephones, modems, and computers. Initially, error correction systems were employed in messages sent by telegraph, where the message was transmitted in Morse code, in which letters were made up of dots and dashes: any distortion would introduce an error in the message. Similarly, today most electronic communication is done digitally, where the message is communicated in bits, which can be recorded either as a 1 or as a 0. Error is introduced by noise in the system that can be caused by faulty equipment, poor connections, electronic noise, and in wireless communication by atmospherics, such as lightning, as the static you hear on the radio or the splashes of dots you see on the TV screen during a storm.

An example of where error correction was of unusual importance was in 1972 when the *Mariner* space probe flew past Mars and transmitted pictures back to Earth. The weak signal received by the spacecraft had to switch on a camera in the satellite. There was no room for error because the pictures had to be taken when the craft was in the right position over Mars. In 1979, the *Voyager* probe began transmitting color pictures of Jupiter. The source alphabet for the pictures taken by the *Mariner* in 1972 consisted of 64 shades of gray, whereas that for the *Voyager* had 4,096 color shades, so the need for correct transmission, and therefore error correction, was all the more important.

Some electronic systems experience more errors than others. Optical disk, for instance, has a higher rate of error than magnetic disk, and magnetic tape has a higher error rate than magnetic disk. On the other hand, fiber optic cable and semiconductor memory have a low error rate. Error rate can be measured by the number of bit errors divided by the total number of bits transferred. With an optical disk one might see about 1 bit error in every 100,000 bits transferred, whereas with magnetic disk there might be 1 error bit for every billion bits transferred. Bit error may also be measured in units of time. If you transfer a billion bits per second, you have a bit error every second. Some disk drives transfer at the rate of 40 million bits per second, so that a bit error occurs every 25 seconds.

Error correction can be performed by hardware or software. As transfer rates are ever increasing and manufacturers are squeezing more bits into storage devices, more errors are bound to occur, thus error correction has to be done by hardware.

How are Latin squares used in producing error correction codes? One might repeat the message several times so that the message received most frequently could be expected to be the correct message. This is a time-consuming procedure, and more efficient methods that employ Latin squares have been devised. These, however, involve a highly technical process and therefore will not be presented here. For those who can handle the math, we suggest chapter 10 of *Applied Combinatorics* [58].

12.3   Combinatorics

Combinatorics, or combinatorial analysis, is a branch of mathematics that, with the advent of high-speed computers, has become very popular in recent years for solving complex, difficult problems in many fields. It deals with permutations and combinations, especially in statistics and probability, and is concerned with the study of arrangements, patterns, designs, schedules, and so on, such as we have been discussing with Latin squares. Statistics and error correction codes might be subsumed under the title of Combinatorics.

In chapter 9 of his book ([58]), Roberts mentions a number of different practical problems that have been handled with combinatorics. These include Latin square arrays for testing tire tread wear, spinning synthetic yarn, determining fuel economy, comparing dishwashing detergents, doing market research, and studying prosthodontics, cardiac drugs, TB in cattle, and the routing of garbage trucks and traffic distribution for the airlines.

12.4   The Future

Basic science usually precedes practical applications of knowledge. The insecticide DDT and the antibiotic penicillin came into practical use during World War II, but both were known many years before. The same is true of many of the drugs we use as medicines. Mathematics is the language of physics, most of which preceded and made way for our great advances in electronics, aviation, and so on. Magic squares, with their long history, have just begun to find their way into our modern life. As of this writing, for example, electronics manufacturer Toshiba has "patented a magic square technology to achieve a 16-million color display" for their LCD panels [75]. This is so new that we do not know how their magic square technology works, but we can be sure this is only a sample of the future for the practical applications of magic squares.

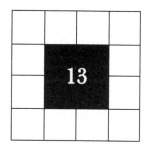

Some Puzzles for You

All solutions to these puzzles appear in Chapter 16.

13.1 Magic Square Puzzles

1. **Magic square with odd numbers.** Make a perfect magic square of order 3, with magic constant 27, made up of only consecutive odd numbers beginning with 1 in any cell. That is, the magic square is made using the numbers 1, 3, 5, 7, 9, 11, 13, 15, 17 (Peterson [54]).

2. **Magic square with even numbers.** Construct a perfect magic square of order 3, with magic constant 30, composed of only consecutive even numbers beginning with 2 in any cell. That is, the magic square is made using the numbers 2, 4, 6, 8, 10, 12, 14, 16, 18 (Peterson [54]).

3. **Magic square with prime numbers.** Prime numbers are numbers that are divisible only by themselves and 1. Make a magic square of order 3, having a magic constant 177, with prime numbers 5, 17, 29, 47, 59, 71, 89, 101, and 113. Start with 5 in any cell (Pickover [56]).

4. **Annihilation magic square.** This should be an easy one. In fact, there is "nothing" to it. Using an order 4 square and numbers 1 to 8 and −1 to −8, find the configuration that will give the sum of the rows, columns, and diagonals to be zero.

5. **Franklin's magic square of order 4.** A magic square of order 4 can be constructed from the sequence of numbers from 1 through 16 as follows (Willis [76]).

   (a) Write the numbers as in the square as shown in Figure 13.1.

   (b) Observe in Figure 13.1 the following properties.

      i. The sum of each number in the first and fourth column equals the addition of the numbers in the second and third columns, for the same row.

1	2	3	4
8	7	6	5
9	10	11	12
16	15	14	13

Figure 13.1   Initial 4 × 4 square.

That is, for the first and fourth the sum gives

$$[1, 8, 9, 16] + [4, 5, 12, 13] = [5, 13, 21, 29],$$
$$(13.1)$$

and for the second and third it is

$$[2, 7, 10, 15] + [3, 6, 11, 14] = [5, 13, 21, 29].$$
$$(13.2)$$

ii. The sum of each number in each column equals the magic constant 34. If any number is moved up or down along the same column, it does not change the result of addition.

iii. The sum of the first and last row equals the sum of the second and third row, which is half the magic sum, that is, 17:

$$[1, 2, 3, 4] + [16, 15, 14, 13] = [17, 17, 17, 17].$$
$$(13.3)$$

and for the second and third rows it is

$$[8, 7, 6, 5] + [9, 10, 11, 12] = [17, 17, 17, 17,].$$
$$(13.4)$$

(c) Interchange the numbers 2 and 3 on the first row with the numbers 15 and 14 on the fourth row on the same column of Figure 13.1. The result of the addition of each number along each column does not change, that is, it is still 34.

(d) Interchange the numbers 7 and 6 on the second row with the numbers 10 and 11 on the third row. It gives the semi-perfect magic square shown in Figure 13.2.

1	15	14	4
8	10	11	5
9	7	6	12
16	2	3	13

Figure 13.2   4 × 4 magic square obtained.

16	3	2	13
5	10	11	8
9	6	7	12
4	15	14	1

Figure 13.3   Dürer-Franklin magic square.

Puzzle. Start with a square similar to but not identical with the square in Figure 13.1 and use the same technique to obtain the Dürer-Franklin 4 × 4 magic square shown in Figure 13.3.

6. **Palindrome magic square of order 3.** Make a magic square of order 3 that is a palindrome (the value in a palindrome is the same read forward or backward) and has a magic sum of 666, using only the digits 1, 2, 3. The numbers must contain three digits, and repetition is allowed. Therefore, the numbers must be greater than 100, such as 121, 111, or 363.

7. **Magic square of order 4.** Many centuries ago in France, Bernard Frénicle de Bessey produced 880 magic squares of order 4, and recent research has shown that 880 is all there are. Try your hand to see how many of them you can reproduce that have the number 1 in the first row and first column.

8. **Composite numbers.** Construct an order 3 magic square comprised of the smallest consecutive composite numbers. Remember that a *composite number* is defined as a number that is neither prime nor one. Hint: the magic sum is 354 (Gardner [16]).

9. **Multiplication magic square of order 3.** Construct a multiplication magic square of order 3 with numbers, the smallest of which is 1 and the largest is 36, giving a magic constant of 216. Figure 13.4 shows a hint for the construction of the magic square.

10. **Fix it.** Given the square in Figure 13.5, verify if it is magic or not. If not, fix it.

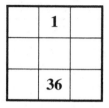

Figure 13.4   Hint for the construction of the multiplication magic square.

11	10	41	60	17
18	49	6	5	24
62	19	13	7	1
2	21	20	14	8
9	3	22	16	52

Figure 13.5   A 5 × 5 square.

13.2   Sudoku Puzzles

		8		4				
9							7	
		7	5		3	1	8	
					2			
	2				8		5	7
	1			9		4		
		9	4		6			
		5	1			9		6

Figure 13.6   Sudoku.

	a	b	c	d	e	f	g	h	i
A		2							1	6
B	1								4	
C	5			9			7			
D			8			3				9
E			2		1					
F		3		5	9					
G		6					4			2
H				7						1
I		4		3		8				

Figure 13.7   Another Sudoku.

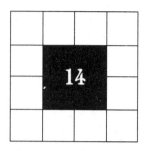

Further Reading

William Schaaf published a list of early books on magic squares [61], which he divided into sections: before 1800; 1800–1880; 1881–1900; and 1901–1940. Frank J. Swetz has an excellent bibliography of fifteen pages in his book, *Legacy of the Luoshu* [71]. Mark Farrar listed a number of references on the Internet ([87]) in which he found old English books on magic squares not on other lists and, in addition, some that are of special interest to professional magicians, who use them in their performances.

However, the most comprehensive survey of the literature is the *Bibliography* of Harvey Heinz. This list takes up eighteen pages

on the Web and begins with sources of information on eighteenth-
and nineteenth-century books as well as titles of some books of
pre-eighteenth-century periods. His list is divided into sections as
follows: books, chapters in books, published papers, papers in the
Journal of Recreational Mathematics, and the literature on magic
stars. Listed alphabetically by author, each mention gives full references plus Heinz's comment on the contents and other useful
material [91].

The *Zen of Magic Squares, Circles, and Stars* by Clifford A. Pickover is a fun book stuffed with interesting material and is unique
for its excellent art.

Paul C. Pasles is a mathematician who discovered Franklin's
long-lost "perfect" 16 magic square. His newly published book, *Benjamin Franklin's Numbers*, shows us just one more side of Franklin's
genius.

Legacy of the Luoshu by Frank J. Swetz gives us a different vision
of magic squares, namely of the square of 3, first reported by the
ancient Chinese and its development shrouded in mysticism.

And a classic in the field, by W. S. Andrews and contributors,
is the encyclopedic book *Magic Squares and Cubes*, which was first
published in 1908 and is still in print.

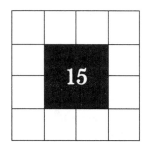

Magic Squares Terminology

Antimagic squares A particular case of heterosquares, where the addition of each number on each row, column, and two diagonals gives numbers that form a consecutive sequence of $2(n+1)$ integers, where n is the order of the square.

Associated magic square A magic square where all pairs of cells diametrically equidistant from the center of the square equal the sum of the first and last terms of the sequence, or $n^2 + 1$ (Heinz [97]). It is also called *symmetrical* or *center-symmetric*.

Bent column A line that starts at the top middle to right of the square, goes down at an angle of 45° until the middle of the left side of the square, then it bends back at an angle of 90° until it reaches the bottom of the square. If the lines are bent left, they have the shape of a less than sign (<) as shown in Figure 10.29. The shape of a greater than (>) sign results if the lines start top middle to right, then are bent to the right, as shown in Figure 10.31.

Bent row A line that starts at the top left side of the square, goes up at an angle of 45° until the middle of the square, then it bends down at angle of 90° until it reaches the right side of the square. If the lines are bent up, they have the shape of a V as shown in Figure 10.25. If the lines are bent down they have the shape of an upside-down V as shown in Figure 10.27.

Bimagic square A magic square that remains magic when each number is replaced with its square (x^2). These squares are also called *double magic*. So far the smallest bimagic square is of order 8.

Bordered magic square A *nested* or *concentric* magic square that has all the usual properties of a simple magic square plus the additional property that if one or more borders are removed the remaining squares are still magic. See Figure 8.13.

Evaluation of the magic constant The magic sum for a $(n \times n)$ square can be evaluated using the expression

$$M_n = \frac{n}{2}(n^2 + 1). \qquad (15.1)$$

For example, the magic constant for the magic squares in Figure 15.1 is evaluated as follows:

$$M_4 = \frac{4}{2}(4^2 + 1) = 2(16 + 1) = 34. \qquad (15.2)$$

The notation M_n is used to represent the magic sum M of a square of size n.

Evaluation of the magic number of a star

To derive the expression (6.33), which is repeated in equation (15.3)

$$M_{star} = 4n + 2, \tag{15.3}$$

note that a standard star with n points contains a polygon with n vertices, which gives a total of $2n$. Let the integers $1, 2, \ldots, 2n$ be distributed one at each of the $2n$ points, such that the sum of each number along each chord line adds up the same value M_{star}.

The sum of the numbers $1, 2, \ldots, 2n$ is given by

$$S_{2n} = \frac{1}{2}(2n + 1)n = n(2n + 1). \tag{15.4}$$

Because there are n chord lines in a star, the sum of all magic constant M_{star} obtained from each chord line equals nM_{star}. Because each point containing an integer is intercepted by two chord lines, the integer on that point is counted twice during the computation of nM_{star}, therefore,

$$nM_{star} = 2S_{2n} = 2n(2n + 1), \tag{15.5}$$

which gives the expression

$$M_{star} = 2(2n + 1) = 4n + 2. \tag{15.6}$$

for the computation of the magic number of a star.

Heterosquare A square with the numbers from 1 to n^2, where the addition of each number on each row, column, and two diagonals gives different numbers.

Latin square A square arrangement of n numbers $1, 2, \ldots, n$ such that each row and column contains the n numbers without repetition in each row or column.

Magic cube A three-dimensional array of numbers where each number along each line on each of the three directions sum to the magic constant.

Magic square

1. *Standard* or *pure magic square:* has the cells filled with consecutive numbers 1 through n^2, arranged in a square form, where n is the size (order) of the square, as shown in Figure 15.1(a), with the property that the sum of the numbers along each row and each column gives the same value, which is called the *magic sum, magic constant,* or *magic number.*

2. *Perfect magic square:* a standard magic square with the addition that the sum of each number on each diagonal also equals the same row/column value, as shown in Figure 15.1(b).

3. *Semi-perfect:* a standard magic square is called semi-perfect, if the sum of each number on each diagonal does not equal the same value obtained for each row and each column, as shown in Figure 15.1(a).

4. *Enumeration of rows and columns:* it is standard practise to enumerate the rows and columns starting at the top left corner of the square.

Magic square cell This is each small square, in a magic square, that contains a number.

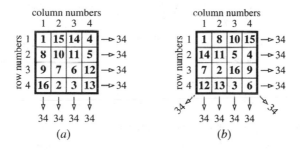

Figure 15.1   (a) 4 × 4 semi-perfect standard magic square; (b) Perfect magic square.

Magic square property A rule applied to n numbers of a magic
square of order n, that gives the magic number.

Magic sum, magic constant, magic number If each number in
each row and in each column adds up to the same value,
this value is called the *magic sum, magic constant,* or *magic
number.* For example, the magic constant of the squares in
Figure 15.1 is 34.

Main diagonal A line from the top left corner of the square to the
bottom right corner. See below for *secondary diagonal.*

Odd and even magic square

- *Odd square:* the size or order of the square is an odd
  number.
- *Even square:* the size or order of the square is an even
  number.

Secondary diagonal The line from the top right side of the square
to the bottom left side of the square.

Size or order of a magic square In a magic square the number n of
rows equals the number of columns. The number n is called
the *size* or *order* of the magic square. The size or order can
be denoted as $n \times n$, where the first n refers to the number
of rows, and the second n refers to the number of columns.
It is also usual to refer to the size of a magic square as *magic
square of order n.*

For example, the size of the magic squares in Figure 15.1
can be denoted as:

- a 4×4 magic square, or
- a fourth-order magic square.

Trimagic square A bimagic square such that when each number
is cubed (x^3), the resultant square is still magic. They are
also referred to as *triple magic.* The triple magic square of
order 12 looks to be the smallest one that exists.

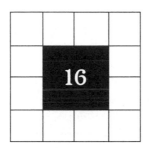

Solutions to the Puzzles

16.1 Magic Square Puzzles

1. **Magic square with odd numbers.**

   A magic square of order 3 contains cells for nine numbers. Nine odd numbers beginning with 1 are 1, 3, 5, 7, 9, 11, 13, 15, and 17. The middle number in this series is 9, and this is also the middle number of the square. It may be attained by adding the first and last numbers of the series and dividing the sum by two. Thus $1 + 17 = 18/2 = 9$. It can also be achieved by adding all the numbers in the series and dividing by nine, thus $81/9 = 9$.

The magic constant is obtained by adding all the numbers in the sequence and dividing by the order of the magic square, thus $81/3 = 27$. An easier way is to multiply the middle term by the order of the square, namely, $9 \times 3 = 27$. Now, knowing the middle number and the magic constant, we can find four triple sets with 9 in the middle that add to give the magic constant.

They are (1, 9, 17), (3, 9, 15), (5, 9, 13), and (7, 9, 11). These sets are then placed strategically in the rows, columns, and diagonals so that the square is magic in all directions, as in Figure 16.1.

2. **Magic square with even numbers.**

The same logic applies here as was used in Puzzle 1. The nine even numbers filling the square are 2, 4, 6, 8,

11	1	15
13	9	5
3	17	7

Figure 16.1 Perfect order 3 magic square with odd numbers.

8	18	4
6	10	14
16	2	12

Figure 16.2 Perfect order 3 magic square with even numbers.

71	89	17
5	59	113
101	29	47

Figure 16.3 Magic square with prime numbers.

10, 12, 14, 16, and 18. The middle number is 10, and is arrived at as in Puzzle 1. The magic constant is the sum of all the numbers divided by three; $90/3 = 30$. The sets of three are (2, 10, 18), (4, 10, 16), (6, 10, 14), and (8, 10, 12). With a little manipulation you will have the magic square (see Figure 16.2).

3. **Magic square with prime numbers.**
   Again the logic used in puzzles 1 and 2 applies. The nine prime numbers to fill the square are 5, 17, 29, 47, 59, 71, 89, 101, and 113. The middle number in the series and the square is 59. The magic constant is $3 \times 59 = 177$. The four sets, (5, 59, 113), (17, 59, 101), (29, 59, 89), and (47, 59, 71) are arranged in the rows, columns, and diagonal as in Figure 16.3.

   To make it harder, have friends try these three puzzles without knowing the magic constant and see how long it takes for them to solve it.

4. **Annihilation magic square.**
   Observe that the sum along each horizontal line in (16.1) equals zero and along the vertical it equals 9 or −9.

$$
\begin{array}{rrrrcl}
1 & -2 & -3 & 4 & = & 0 \\
+ \quad 8 & -7 & -6 & 5 & = & 0 \\
\hline
9 & -9 & -9 & 9 & &
\end{array}
\qquad (16.1)
$$

If the signs in (16.1) are interchanged, that is, making the positive negative and vice versa, it gives the arrangement in (16.2):

$$
\begin{array}{r}
-1 \quad 2 \quad 3 \quad -4 \ = \ 0 \\
+ \quad -8 \quad 7 \quad 6 \quad -5 \ = \ 0 \\
\hline
-9 \quad 9 \quad 9 \quad -9
\end{array}
\qquad (16.2)
$$

where the sum along the horizontal equals zero, and along the vertical it is -9 or 9.

Note that the addition of two consecutive numbers n and $n+1$ where one is positive and the other negative will always give $+1$ or -1, that is $-n+(n+1) = 1$ and $n+[-(n+1)] = -1$. You should test a few examples to verify this property. As a consequence, the sum along the horizontal in (16.1) and (16.2) will always be zero.

If the numbers in (16.1) and (16.2) are put together, then it follows

$$
\begin{array}{r}
1 \quad -2 \quad -3 \quad 4 \ = \ 0 \\
+ \quad 8 \quad -7 \quad -6 \quad 5 \ = \ 0 \\
-1 \quad 2 \quad 3 \quad -4 \ = \ 0 \\
-8 \quad 7 \quad 6 \quad -5 \ = \ 0 \\
\hline
0 \quad 0 \quad 0 \quad 0
\end{array}
\qquad (16.3)
$$

which has the property that the sum along the horizontal or vertical is zero. Note that due to the construction in (16.1) and (16.2), there is no repeated number in (16.3), because -4 and 4 for example, are different numbers. Therefore, the numbers in (16.3) form an annihilation magic square, and they are shown in Figure 16.4 in the magic square format. It is not a perfect magic square because the diagonals do not annihilate.

A second example is shown in Figure 16.5.

1	-2	-3	4
8	-7	-6	5
-1	2	3	-4
-8	7	6	-5

Figure 16.4 A 4 × 4 annihilation magic square.

8	-7	-6	5
-4	3	2	-1
1	-2	-3	4
-5	6	7	-8

Figure 16.5 A 4 × 4 annihilation magic square.

The numbers in (16.1) and (16.2) can also be arranged as

$$\begin{array}{r} 8 \quad -7 \quad -6 \quad 5 = 0 \\ + \quad -4 \quad 3 \quad 2 \quad -1 = 0 \\ \hline 4 \quad -4 \quad -4 \quad 4 \end{array} \qquad (16.4)$$

which was used to generate the first and second row of the square, and

$$\begin{array}{r} 1 \quad -2 \quad -3 \quad 4 = 0 \\ + \quad -5 \quad 6 \quad 7 \quad -8 = 0 \\ \hline -4 \quad 4 \quad 4 \quad -4 \end{array} \qquad (16.5)$$

which gives the third and fourth rows. Because the numbers were arranged such that the sum of the columns gives numbers with opposite signs, which

4	3	2	1
5	6	7	8
12	11	10	9
13	14	15	16

Figure 16.6   Initial square.

4	3	2	1
5	6	7	8
12	10	11	9
13	15	14	16

Figure 16.7   Interchanging 10, 15 with 11, 14.

implies that their sum will be zero. Figure 16.5 shows the resultant annihilation square, which is perfect.

The annihilation shown in Figure 16.5 can also be found in Moscovich [51].

5. **Franklin's magic square of order 4.**

Write the square in the form shown in Figure 16.6, that is, start at the first row, fourth column with the number 1, then move to the second, third, and fourth row back and forth.

Interchange 10 and 15 on the second column with 11 and 14 on the third column. The result of the sum of each number along the second and third columns does not change because $10 + 15 = 11 + 14 = 25$. The new square is shown in Figure 16.7.

4	3	2	1
5	10	11	8
12	6	7	9
13	15	14	16

Figure 16.8 Interchanging 6, 7 with 10, 11.

4	3	2	1
5	10	11	8
9	6	7	12
16	15	14	13

Figure 16.9 Interchanging 9, 16 with 12, 13.

Interchange 6, 7 on the second row with 10, 11 on the third row. They are on the same columns; the result, 34, of the sum of each number along each of these columns does not change. The new square is shown in Figure 16.8.

Interchange 9 and 16 on the fourth column with 12 and 13 on the first column. This does not change the sum of each number on each column; $9 + 16 - 12 + 13 = 25$. See Figure 16.9.

The last step is to interchange 4 with 16 on the first column, and 1 with 13 on the fourth column. This transformation does not change the sum of each number long each column.

Benjamin Franklin's square is shown in Figure 16.10.

16	3	2	13
5	10	11	8
9	6	7	12
4	15	14	1

Figure 16.10   Benjamin Franklin 4 × 4 magic square.

232	313	121
111	222	333
323	131	212

Figure 16.11   Palindrome magic square.

6. **Palindrome magic square of order 3.**

   In the solution of Puzzle 1 it was shown that in a 3 × 3 magic square, the middle number multiplied by the size of the square equals the magic constant. Therefore, the middle number is $666/3 = 222$. As a consequence the sum of the other two numbers on the second row equals $666 - 222 = 444$. A trivial choice is to split 444 into two palindromic numbers such as $111 + 333 = 444$.

   First column: The sum of the number on the first and third cell is equal to $666 - 111 = 555$. Therefore, only the digits 2, 3, where $2 + 3 = 5$, can be used. The solution is 232 and 323, which gives $232 + 323 = 555$.

   Second column: The case is similar to the one of the first column. The two numbers must add up $666 - 222 = 444$, that is, only the digits 1 and 3 can be used, which gives the solution 131 and 313.

Third column: The two numbers must add up 666 −
333 = 333, that is, only the digits 1 and 2 can be used,
which gives the solution 121 and 212.

The numbers on the four corners must be chosen
such that the the sum of each number on each diagonal also equals the magic constant. Because the two
numbers for the second column are independent of the
diagonals, they can be chosen, for example, as 313 on
the first row and 131 on the third row.

The diagonal contains the number 222, therefore,
the other two numbers must add up to 444, which
implies the decomposition 4 = 1 + 3 and 4 = 2 + 2.
Then the following possible pairs of numbers (313,
131), (121, 323), and (232, 212).

If the number 232 is chosen for the first cell of the
first column, then the first cell of the third column must
contain the other number of the pair, that is, 121. The
number 232 on the first cell of the first column implies
that the number on the third cell on the same column
must be 323. Analogously, the third number on the
third column must be 212. Then you have the magic
square in Figure 16.11.

7. **Magic square of order 4.**
   Consult De Haas [12] and count them; you will find
   26 × 8 = 208, that is, 26 rows with 8 squares on each
   row that have the number 1 in the first row and first
   column.

8. **Composite numbers.**
   A *composite number* is defined as one that is neither
   prime nor one. The first step to solve the problem is to
   have (or construct) a table of all prime numbers less
   than 300 (see Table 16.1).

   A 3 × 3 standard magic square contains the numbers 1, 2, . . . , 9. Therefore, seek in the table a prime
   number such that, by adding 9 to it, the result will be

TABLE 16.1 Prime numbers less than 300.

2	3	5	7	11	13	17	19	23	29
31	37	41	43	47	53	59	61	67	71
73	79	83	89	97	101	103	107	109	113
127	131	137	139	149	151	157	163	167	173
179	181	191	193	197	199	211	223	227	229
233	239	241	251	257	263	269	271	277	281
283	293								

smaller than the next prime number in the table. This guarantees that the sum of any number from 1 to 9 to the prime number will never be prime. For example, $43 + 9 = 52$, where 52 is greater than the next prime number after 43, which is $43 + 4 = 47$. The smallest prime number that provides a solution is 113, that is, $113 + 9 = 122$, which is smaller that the next prime number, 127.

Then take any standard 3 × 3 magic square and add 113 to each of the numbers, and you will have a 3 × 3 with composite numbers. For example, Figure 16.12(a) shows a standard 3 × 3 magic square and Figure 16.12(b) shows the composite square.

The magic constant of the composite magic square in Figure 16.12 equals the magic constant of a 3 × 3, which is 15, magic square plus 3 × 113. For example, the addition of each number along the first row gives

$$M = (4 + 113) + (3 + 113) + (8 + 113)$$
$$= (4 + 3 + 8) + 3 \times 113 = 15 + 3 \times 113 = 354$$

$$\text{(16.6)}$$

Figure 16.13(a) shows another standard 3 × 3 magic square and Figure 16.12(b) shows the composite square

4	3	8
9	5	1
2	7	6

(a)

117	116	121
122	118	114
115	120	119

(b)

Figure 16.12   (a) A standard 3 × 3 magic square; (b) A composite 3 × 3 magic square, obtained by adding 113 to each number of the standard magic square in (a).

6	1	8
7	5	3
2	9	4

(a)

187	182	189
188	186	184
183	190	185

(b)

Figure 16.13   (a) A standard 3 × 3 magic square; (b) A composite 3 × 3 magic square, obtained by adding 181 to each number of the standard magic square in (a).

obtained by adding 181 to the numbers on each cell. Observe that $181 + 9 = 190$, and the next prime number after 181 is 191, therefore, the addition of each numbers $1, 2, \ldots, 9$ to 181 will never give a prime number.

Note that the number 187 is not prime since $11 \times 17 = 187$.

The magic constant for the composite magic square in Figure 16.13 is $M = 15 + 3 \times 181 = 558$.

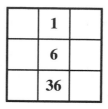

Figure 16.14   Evaluation of the number to fill the middle cell.

There are many other prime numbers satisfying the condition that when 9 is added to it, the sum is smaller than the next prime number, for example, 139, 199, 241, and so on.

9. **Multiplication magic square of order 3.**

The second column gives $1 \times x \times 36 = 216$, therefore, $x = 216/6 = 36$. Introducing this value on the middle cell, the square becomes what is shown in Figure 16.14.

To obtain the numbers for the third row, divide $216 \div 36 = 6$. Because $6 = 2 \times 3$, it gives two solutions for the last row, namely, (2, 36, 3) or (3, 36, 2). Try (3, 36, 2).

For the second row, the product of the unknown terms is $216 \div 6 = 36$. The numbers with a product equal to 36 are 2×18, 3×12, 4×9. Because 2 and 3 have already been used, the products 2×18, 3×12 are eliminated. The product 4×9 gives the solution for the second row, (4, 6, 9) or (9, 6, 4). Trying (4, 6, 9) it gives the following square in Figure 16.15.

To obtain the numbers for the first row, it is easier to work with the first and third column. For the first column, it gives $3 \times 4 \times y = 216$, therefore, $y = 18$. For the last column, $2 \times 9 \times w = 216$, which gives $w = 12$. Because 18 and 12 have not yet been used, they are the solution. Using 18 for the first row and 12 for the last gives the magic square in Figure 16.16.

	1	
4	6	9
3	36	2

Figure 16.15 Numbers filling the second and third rows.

18	1	12
4	6	9
3	36	2

Figure 16.16 Multiplication magic square.

10. **Fix it.**

First observe that a standard 5×5 magic square contains the numbers $1, 2, \ldots, 25$, and the given square has five numbers greater than 25, which are 41, 60, 49, 62, and 52.

Recall that the magic number of a standard magic square is given by

$$M_5 = \frac{n}{2}(n^2 + 1) = \frac{5}{2}(5^2 + 1) = 65. \qquad (16.7)$$

Because the numbers in the fourth row are in $1, 2, \ldots, 25$, this row may be correct. To verify, evaluate the sum of each number, that is,

$$2 + 21 + 20 + 14 + 8 = 65. \qquad (16.8)$$

which equals the magic sum.

The addition of the numbers along each column will not give the magic sum, because each row has a number greater than 25. There is only one number in each column above 25, so it is possible to evaluate what would be the right number that will give the magic sum. Call this number x, then for the first column

$$11 + 18 + x + 2 + 9 = 65, \qquad (\mathbf{16.9})$$

which gives $x = 65 - (11 + 18 + 2 + 9) = 25$. Repeating the same process for the other columns, it follows for the second column

$$10 + x + 19 + 21 + 3 = 65; \qquad (\mathbf{16.10})$$

then, $x = 65 - (10 + 19 + 21 + 3) = 12$.

For the third column,

$$x + 6 + 13 + 20 + 22 = 65; \qquad (\mathbf{16.11})$$

then, $x = 65 - (6 + 13 + 20 + 22) = 4$. For the fourth column,

$$x + 5 + 7 + 14 + 16 = 65; \qquad (\mathbf{16.12})$$

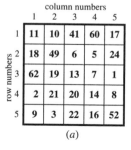

Figure 16.17   (a) Given 5×5 square; (b) 5×5 magic square.

then, $x = 65 - (5 + 7 + 14 + 16) = 23$. For the fifth column,

$$17 + 24 + 1 + 8 + x = 65; \qquad (16.13)$$

then, $x = 65 - (17 + 24 + 1 + 8) = 15$.

Verification: evaluation of the magic sum along the first row

$$11 + 10 + 4 + 23 + 17 = 65. \qquad (16.14)$$

The resultant magic square is shown in Figure 16.17. Check the two diagonals to see if this is a perfect magic square. The square used in this puzzle can be found on Suzuki's Web site [112].

16.2   Sudoku Puzzles

1	5	8	7	4	9	2	6	3
9	3	2	8	6	1	5	7	4
4	6	7	5	2	3	1	8	9
5	9	3	6	7	2	8	4	1
6	2	4	9	1	8	3	5	7
8	7	1	3	5	4	6	9	2
7	1	6	2	9	5	4	3	8
2	8	9	4	3	6	7	1	5
3	4	5	1	8	7	9	2	6

Figure 16.18   Solution for Figure 13.6.

	a	b	c	d	e	f	g	h	i
	3	**2**	9	4	8	7	5	**1**	**6**
	1	7	6	2	3	5	9	**4**	8
	5	8	4	**9**	6	1	**7**	2	3
	4	1	**8**	6	7	**3**	2	5	**9**
	9	5	**2**	8	**1**	4	3	6	7
	6	**3**	7	**5**	**9**	2	1	8	4
	8	**6**	3	1	5	9	**4**	7	**2**
	2	9	5	**7**	4	6	8	3	**1**
	7	**4**	1	**3**	2	**8**	6	9	5

Figure 16.19   Solution for Figure 13.7.

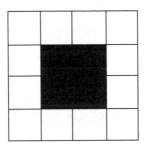

References

[1] Abbott, Edwin, *Flatland: A Romance of Many Dimensions*. Seely & Co., 1884.

[2] Andrews, William Symes, *Magic Squares and Cubes*. Second edition. Dover Publications, New York, 1960.

[3] Ibid., p. 172.

[4] Ibid., p. 171.

[5] Ibid., p. 241–42.

[6] Bragdon, Claude. *The Frozen Fountain*. Alfred A. Knopf, New York, 1932.

[7] Brandreth, Gyles. *Number Play*. Rawson Associates, 1984, pp. 68–69.

[8] Cammann, Schuyler. "Islamic and Indian Magic Squares," I and II. *History of Religions* 8 (3), 8, 1969.

[9] Ibid., pp. 27–29.

[10] Davis, Bob. *An Introduction to Agricultural Statistics*. Delmar, Albany, NY, 2000, pp. 138–41.

[11] Descombes, René. *Les Carrés magiques: Histoire, théorie et technique du carré magique, de l'antiquité aux recherches actuelles*. Second edition. Librarie Vuibert, Paris, 2000.

[12] de Haas, K. H., *Frénicle Index*. D. van Sijn & Zonen, Rotterdam-Centrum Printer & Publishers, Rotterdam, Holland, 1935.

[13] Denes, I., and A. R. Keedwell. *Latin Squares and Their Applications*. Academic Press, New York, 1974, pp. 194–229.

[14] Dudeny, H. E. *Encyclopedia Britannica*, vol. 15, 1954, pp. 627–30.

[15] Emanouilidis, Emanuel. "Concatenation on Magic Squares." *Journal of Recreational Mathematics* 31 (2), 110–11 (2002–2003).

[16] Gardner, Martin. "Mathematical Games." *Scientific American* 234 (1), 118–23, 1976.

[17] Gridgeman, N. T. "Magic Squares Embedded in a Latin Square." *Journal of Recreational Mathematics*, 5 (4), 1972.

[18] Hayes, Brian. "Unwed Numbers: The Mathematics of Sudoku, a Puzzle that Boasts No Math Required." *American Scientist* 94 (1), 12, 1996.

[19] Heinz, Harvey D., and John Robert Hendricks. *Magic Square Lexicon Illustrated*, H. D. Heinz, Toronto, 2000, pp. 18–19.

[20] Ibid., pp. 152–53.

[21] Ibid., pp. 37–38.

[22] Ibid., p. 56.

[23] Ibid., p. 4.

[24] Ibid., p. 142.

[25] Ibid., p. 69.

[26] Ibid., p. 14.

[27] Hendricks, John Robert. "The Third-Order Magic Cube Complete." *Journal of Recreational Mathematics* 5 (1), 43–50, 1972.

[28] Hendricks, John Robert. "The Pan-3-Agonal Magic Cube." *Journal of Recreational Mathematics* 5 (1), 51–52, 1972.

[29] Hendricks, John Robert. "Inlaid Magic Squares." *The Manitoba* (Canada) *Mathematics Teacher*, 17 (4), May 1989.

[30] Holmes, R. "The Magic Magic Square." *Mathematical Gazette*, 54 (390), 376, December 1970.

[31] Horner, W. W. "Addition Multiplication Magic Square of Order 8." *Scripta Math* 21, 23–27, 1955.

[32] Johnson, Allan W. Jr. "Palindromes and Magic Squares." *Journal of Recreational Mathematics*, 21 (2), 97–100, 1989.

[33] Johnson, Allan W. Jr. "A Square Magic Square." *Journal of Recreational Mathematics*, 22 (1), 38, 1990.

[34] Kurchan, Rodolfo Marcelo. "Solution to Problems and Conjectures." *Journal of Recreational Mathematics*, 23 (1), 69, 1991.

[35] Labaree, Leonard W., ed., and Whitfield J. Bell Jr., associate ed. *The Papers of Benjamin Franklin: January 1, 1745 through June 30, 1750, 3*, 458–9. Yale University Press, New Haven, CT, 1961.

[36] Labaree, Leonard W., ed., and Whitfield J. Bell Jr., associate ed. *The Papers of Benjamin Franklin: July 1, 1750 through June 30, 1753, 4, 7*. Yale University Press, New Haven, CT, 1961.

[37] Ibid., pp. 392–403.

[38] Labaree, Leonard W., Ed. and Whitfield J. Bell Jr., associate ed. *The Papers of Benjamin Franklin: January 1, through December 31, 1765, 12*, p. 148, Yale University Press, New Haven, CT, 1968.

[39] Ibid., p. 148.

[40] Lancaster, Ronald J. "Infinite Magic Squares." *Journal of Recreational Mathematics*, 9 (2), 86–93, 1976–77.

[41] Larson, Harold D. *Magic Squares, Circles, Stars: Grade 5*. 16 pp. booklet published in 1956.

[42] Lemay, J. A. Leo. "The Autobiography." In *Benjamin Franklin's Writings*, p. 1420. Library of America, New York, 1987.

[43] Lindon, J. A. "Anti-Magic Squares." *Recreational Mathematics Magazine* 7, 16–19, February 1962.

[44] Lister, Rodney. *Steps through the Maze: Image, Reflection, Shadow and Aspects of Magic Squares in the Works of Sir Peter Maxwell Davies*. PhD dissertation, Brandeis University, 2001.

[45] Lister, Rodney. "Peter Maxwell Davies Naxos Quartets." *Tempo* 59 (232), 2–12, 2005.

[46] Madachy, Joseph S. *Madachy's Mathematical Recreation*. Dover, New York, 1979.

[47] Ibid., p. 93.

[48] Ibid., pp. 100–101.

[49] Mannke, William J. "A Magic Square." *Journal of Recreational Mathematics*, 1 (3), p. 139, July 1968.

[50] Michell, John. *The View over Atlantis*. Garnstone Press, London, 1969.

[51] Moscovich, Ivan, *Fiendishly Difficult Math Puzzles*. Sterling, New York, 1986, p. 18.

[52] Ondrejka, Rudolf. "666 Again," letters to the editor. *Journal of Recreational Mathematics*, 16 (2), 121, 1983–84.

[53] Patterson, D. D. *An Introduction to Agricultural Statistics*. McGraw-Hill, New York, 1939, pp. 168–178.

[54] Peterson, Ivars. *Mathematical Treks: From Surreal Numbers to Magic Circles*. Mathematical Association of America, 2002, p. 323.

[55] Phillips, J. P. N. "The Use of Magic Squares for Balancing and Assessing Order Effects in Some Analysis of Variance Design." *Applied Statistics*, 13, pp. 67–73, 1964.

[56] Pickover, Clifford A. *The Zen of Magic Squares, Circles, and Stars: An Exhibition of Surprising Structures across Dimensions*. Princeton University Press, Princeton NJ, 2002, pp. 191–192.

[57] Ibid., p. 228.

[58] Roberts, Fred S. *Applied Combinatorics*. Prentice Hall, New Jersey, 1984.

[59] Sadie, Stanley, and John Tyrrell, eds. *The New Grove Dictionary of Music and Musicians*. Oxford University Press, New York, 2003.

[60] Sayles, H. A. "Serrated Magic Squares." In Andrews [2], pp. 241–44.

[61] Shaaf, William L. "Early Books on Magic Squares." *Journal of Recreational Mathematics*, 16 (1), pp. 1–6, 1983–84.

[62] Schimmel, Annemarie. *The Mystery of Numbers*. Oxford University Press, NY, 1993.

[63] Shineman, Edward W. Jr. "Magic Rectangle." *Journal of Recreational Mathematics*, 30 (2), 111, 1999–2002.

[64] Shineman, Edward W. Jr. "Magic L(ightning)." *Journal of Recreational Mathematics*, 31 (3), 167, 2002–2003.

[65] Spencer, Donald D. *Computers in Number Theory*. Computer Science Press, 1982, pp. 185–86.

[66] Stewart, Ian. "Mathematical Recreations." *Scientific American,* 106, January 1997.

[67] Stifel, Michael. *Arithmetica Integra*. Nurenberg, 1544.

[68] Swetz, Frank J. *Legacy of the Luoshu*. Open Court, Chicago, 2002.

[69] Ibid., p. 9.

[70] Ibid., p. 130.

[71] Ibid., pp. 193–207.

[72] Ibid., p. 383.

[73] Tavares, Santiago A. *Generation of Multivariate Hermite Interpolating Polynomials*. Chapman & Hall/CRC, Taylor & Francis Group, Boca Raton, FL, 2006, p. 175.

[74] Ibid., p. 177.

[75] "Controller Support 2 LCD Displays. Toshiba Releases Newly Developed VGA LCD Controller Based on MDDT Technology for 3G Headsets." *Product News Network,* March 8, 2006.

[76] Willis, John. *Easy Methods of Constructing the Various Types of Magic Squares and Magic Cubes with Symmetric Designs Founded Thereon*. Percy Lund, Huaphries & Co, Country Press, Bradford and London, 1909.

[77] Age of Puzzles. http://www.ageofpuzzles.com/Collections/PuzzleHits .htm.

[78] Composed by John Cormie, *The Anti-Magic Square Project*, July 1999. http://io.uwinnipeg.ca/\%7Evlinek/jcormie/index.html.

[79] Davies, Peter Maxwell. *Pasadena Star-News* interview. http://www .angelfire.com/music2/davidbundler/maxwelldavies.html.

[80] Decombes, René. *Les Carrés Magiques: Oeuvres d'Art Associées aux Carrés Magiques.* http://www.kandaki.com/CM-media.php?cat= 1&aut = 6.

[81] Dimond, Jonathan. *Magic Squares.* http://jonathandimond.com/ downloadables/Magic%20Squares.pdf.

[82] Dimond, Jonathan. *Hand Percussion, World & Ethnic Music, Drum Pro World.* http://www.cdipublications.com/media/DP_World_Fall 2003.pdf.

[83] Dürer, Albrecht. *Melencolia I*, 1514. Engraving. Handbook of the collection Herbert F. Johnson *Museum of Art, Cornell University.* http://www.museum.cornell.edu/HFJ/handbook/hb107.html.

[84] Rickey, V. Frederick. *Dürer's Magic Square, Cardanos Rings, Prince Ruperts Cube, and Other Neat Things.* http://www.math.usma.edu/people/Rickey/papers/ShortCourseAlbuquerque.pdf.

[85] Finklestein, David R. *The Relativity of Albert Dürer. Melencolia I.I.* January 10, 2007. http://www.physics.gatech.edu/people/faculty/finkel stein/DurerCode050524.pdf; http://www.physics.gatech.edu/people/faculty/dfinkelstein.html.

[86] Escher, M. C. *Picture gallery "Symmetry."* http://www.mcescher.com/Gallery/gallery-symmetry.htm.

[87] Farrar, Mark S. *Magic Squares—References.* http://www.markfarrar.co.uk/msqref01.htm.

[88] *The Flower of Life.* http://www.world-mysteries.com/sar_sagel.htm; http://en.wikipedia.org/wiki/Image:Flower-of-Life-small.png.

[89] Goudreau Museum of Mathematics in Art and Science, New Hyde Park, NY. http://www.mathmuseum.org/.

[90] Geest, Patrick De. *Recreational Topics from the WORLD! OF NUMBERS.* http://www.worldofnumbers.com/.

[91] Heinz, Harvey D. *Magic Star Definition.* 2000 (This page last updated March 11, 2006). http://www.geocities.com/~harveyh/magicstar_def.htm; http://www.geocities.com/~harveyh/order5.htm.

[92] Ibid., http://www.geocities.com/~harveyh/magicstar.htm.

[93] Ibid., http://www.geocities.com/~harveyh/3-d_star.htm.

[94] Heinz, Harvey D. *Unusual Magic Squares.* http://www.geocities.com/~harveyh/unususqr.htm.

[95] Heinz, Harvey D. *Magic Squares: Four Plus Five Equals Nine.* http://www.geocities.com/~harveyh/magicsquare.htm.

[96] Heinz, Harvey D. *More Magic Squares.* http://www.geocities.com/~harveyh/moremsqrs.htm.

[97] Heinz, Harvey D. *Glossary.* http://www.geocities.com/~harveyh/glossary.htm.

[98] Hendricks, John Robert, *History of the Magic Tesseract.* http://members.shaw.ca/johnhendricksmath/tesseracts.htm.

[99] Hinton, Charles Howard. *Charles Howard Hinton*. January 10, 2007. http://en.wikipedia.org/wiki/Charles_Howard_Hinton.

[100] Ireland, Patric. *Three-Dimensional Magic Square*. http://www.eatornat .net/artists_ireland_sub1_magicsq.htm.

[101] Bayer, Barbara. *Japan Times*. July 3, 2006. http://www.nikoli.co.jp/en/ misc/20060703the_japan_time.htm. Article reprint in Nikoli Web site, with permission from *The Japan Times*.

[102] McNeill, David. "Maki Kaji: First He Gave Us Sudoku." *Independent*, May 2007. http://news.independent.co.uk/media/article 2502489.ece.

[103] Jelliss, G. P. *Classic Mathematical Recreations: Magic Rectangles*. 2002 (updated May 2003). http://www.gpj.connectfree.co.uk/mrm.htm.

[104] *The Largest Magic Square*. http://www.recordholders.org/en/records/ magic.html.

[105] Mochalov, Leonid. *Puzzles of Leonid Mochalov*. http://puzzlemochalov .com/index.htm.

[106] Nakamura, Mitsutoshi, *Magic Rectangles*, 2004–2006 (updated on September 18, 2006). http://homepage2.nifty.com/googol/magcube/ en/rectangles.htm.

[107] Rivera, Carlos. *Prime Puzzles & Problems*. September 1998. http://www.primepuzzles.net.

[108] Trenkler, Marián. http://www.geocities.com/CapeCanaveral/Launch pad/4057/trenkler.htm.

[109] Lyapunov Fractals. *Gallery of Softology: The Science of Software*. http://softology.com.au/gallery/lyapunov01.jpg.

[110] Smirnov, Dmitri N. *Two Magic Squares*. http://www.sibeliusmusic .com/cgi-bin/show_score.pl?scoreid=44843.

[111] Skyone. *The World's Largest Sudoku Puzzle: Win £5000*. http://www .skyone.co.uk/programme/pgcfcature.aspx?pid=48&fid=129.

[112] Mutsumi Suzuki Web site. *Examples of Magic Squares from 3 × 3 through 20 × 20 (by Tamori's Method)*. http://mathforum.org/te/exchange/ hosted/suzuki/MagicSquare.byTAMORI.html.

[113] *Variations of Sudoku*. http://www.sudoku-electronic-puzzles.com/varia tions-on-sudoku-games.htm.

[114] Wikipedia and Answer.com. *Pyramid.* http://en.wikipedia.org/wiki/
Pyramid_%28geometry%29; http://www.answers.com/topic/pyramid-
geometry.

[115] Wikipedia. *Perfect Magic Cube.* http://en.wikipedia.org/wiki/Perfect_
magic_cube.

[116] Wilke, Robert C. *Nested Magic Squares.* http://members.aol.com/
.robertw653/magicsqr.html.

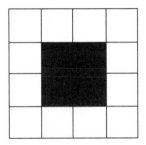

Index